Readings in
Decision Analysis

11·50

OTHER STATISTICS TEXTS FROM
CHAPMAN AND HALL

Problem Solving: A Statistician's Guide
C. Chatfield

Statistics for Technology
C. Chatfield

The Analysis of Time Series
C. Chatfield

Introduction to Multivariate Analysis
C. Chatfield and A. J. Collins

Applied Statistics
D. R. Cox and E. J. Snell

An Introduction to Statistical Modelling
A. J. Dobson

Introduction to Optimization Methods and their Application in Statistics
B. S. Everitt

Multivariate Statistics – A Practical Approach
B. Flury and H. Riedwyl

Multivariate Analysis of Variance and Repeated Measures
D. J. Hand and C. C. Taylor

Multivariate Statistics Methods – A Primer
Bryan F. Manley

Statistical Methods in Agriculture and Experimental Biology
R. Mead and R. N. Curnow

Elements of Simulation
B. J. T. Morgan

Probability: Methods and Measurement
A. O'Hagan

Essential Statistics
D. G. Rees

Foundations of Statistics
D. G. Rees

Applied Statistics: A Handbook of BMDP Analyses
E. J. Snell

Point Process Models with Applications to Safety and Reliability
W. A. Thompson, Jr

Elementary Applications of Probability Theory
H. C. Tuckwell

Intermediate Statistical Methods
G. B. Wetherill

Further information on the complete range of Chapman and Hall
statistics books is available from the publishers.

Readings in Decision Analysis

A collection of edited readings, with accompanying notes, taken from publications of the Operational Research Society of Great Britain.

Simon French

University of Manchester

London   New York
CHAPMAN AND HALL

First published in 1989 by Chapman and Hall Ltd
11 New Fetter Lane, London EC4P 4EE
Published in the USA by Chapman and Hall
29 West 35th Street, New York NY 10001

Phototypesetting by Thomson Press (India) Limited, New Delhi
Printed in Great Britain by St Edmundsbury Press Limited, Bury St Edmunds, Suffolk

ISBN 0 412 31120 8 (hardback) 0 412 32170 X (paperback)

British Library Cataloguing in Publication Data

French, Simon, *1950–*
  Readings in decision analysis
  1. Decision analysis, Mathematical
  models I. Title
  658.4'0354'0724

ISBN 0 412 31120 8
ISBN 0 412 32170 X Pbk

Library of Congress Cataloging in Publication Data

Readings in decision analysis: a collection of edited readings, with accompany-
  ing notes, taken from publications of the Operational Research Society of
  Great Britain/[edited by] Simon French.
    p. cm.
  Bibliography: p.
  Includes index.
  ISBN 0 412 31120 8. ISBN 0 412 32170 X (pbk.)
  1. Decision-making. I. French, Simon, 1950– . II. Operational
Research Society (Great Britain)
T57.95.R39 1989
658.4'03—dc 19   88-25700
                                                        CIP

Contents

vi   Contents

Preface

This collection of readings and notes was developed originally for a tutorial day on Decision Analysis, which I gave in the Spring of 1986 at the behest of the Operational Research Society. It is a feature of the Society's tutorial days that participants are provided with substantial handouts. I have an aversion to writing what has already been written. Thus I developed much of the handout from material previously published in the *Journal of the Operational Research Society*, formerly known as the *Operational Research Quarterly*. The Resulting collection of readings and accompanying notes received favourable comments both from the participants at the event and from several colleagues. I also used the notes with several groups of MSc students taking management options, and again received favourable comment. These comments, and my own ambition, have led me to seek publication in a more permanent form.

I believe that this collection can help satisfy various needs. Firstly, it can serve as a primary text for students on undergraduate and postgraduate management courses. Although decision analysis is fundamentally a quantitative technique, the presentation here involves little beyond basic arithmetic and the ability to read the occasional symbol, perhaps complicated by a subscript or two. Secondly, for students meeting decision analysis on more quantitative courses, e.g. operational research, statistics or mathematics, the collection provides supplementary reading to support more mathematical texts such as my *Decision Theory: An introduction to the mathematics of rationality* (Ellis Horwood, 1986) or Jim Smith's companion volume to this (*Decision Analysis: A Bayesian approach*, Chapman and Hall, 1988). Such supplementary reading is essential if students are not to be left with a 'recipe book' view of decision analysis. They need to see case studies so that they may appreciate how quantitative results are interpreted qualitatively in practice. Thirdly, portable, powerful, interactive computing means that decision analysis is finding more and more applications in industry, commerce and government. Operational research practitioners and management consultants need texts to refresh and update their memories of techniques once only encountered in academic courses.

I hope that what follows is not merely a collection of reprints with a few accompanying notes and comments. I have tried to edit and weave the several extracts and papers together with material of my own so that it forms a coherent whole: a picture of decision analysis as it is practised today. One point that I should make is that, although I have drawn heavily on the writings of others, I do not claim necessarily to have reflected their views fairly. My aim has been to describe decision analysis as I see it. In selecting material to include, in quoting from some papers and in referring to others, I have emphasized the issues that I see as central. I have often discarded or discounted material that I see as peripheral; doubtless, the original authors did not see it as such. If I have offended anyone by doing so, I apologize.

Naturally, I am grateful to many people for assistance:

1. The Operational Research Society for inviting me to give the tutorial day. Indeed, they were foolish enough to ask me to repeat the exercise in February 1988, thus giving me the opportunity of refining my notes again as I was preparing this book.
2. Ray Showell, who has handled the Society's interests in producing this collection in a manner that could not been more helpful.
3. Elizabeth Johnston of Chapman and Hall, who could have found more cooperative, punctual authors elsewhere, but put up with me nevertheless.
4. Peter Hall, Roger Hartley, Patrick Humphreys, Dennis Lindley, Larry Phillips, Jim Smith, Lyn Thomas, Marilena Vassiloglou, Doug White and many others with whom I have discussed the role and purpose of decision analysis.
5. Those authors whose work I have remorselessly pillaged to serve my own ends.

And last, but certainly not least:

6. My family, who nodded knowingly a few years ago when I promised never to write another book, yet supported me wholeheartedly when I did.

To all, thank you.

Simon French,
April 1988   University of Manchester

Articles and extracts from the *Journal of the Operational Research Society* are reproduced by kind permission of Pergamon Press and the Operational Research Society of Great Britain.

Foreword

This book sprang from a Tutorial Event on Decision Analysis conducted by Simon French, which was organized by the Operational Research Society in 1986 during my term of office as President of the Society. Indeed, it was the first such event in a series which is now a well-established feature of the British Operational Research scene.

It is certainly extremely satisfactory that Professor French's path-breaking tutorial has also given us this permanent contribution to the literature on decision analysis. The unusual format, in which case studies and other previous papers illuminate and complement the original text, is one which other authors might well emulate. Indeed, the republication in readily accessible form of so many excellent practical studies would itself be an event to celebrate in any field.

This book, of course, does far more than that. It provides and admirably simple and lucid introduction to the subject of decision analysis, and takes the reader with the minimum of mathematical apparatus through the main concepts and types of application. The reader will reach the boundaries of the subject, where issues are still being hotly debated and new territory staked out. The emphasis of the book, on generating understanding rather than optimal solutions, is one I wholeheartedly support.

Decision analysis is a topic of growing practical importance. It deserves to be taken seriously; so I will take this opportunity to touch briefly on a few questions, some explicitly raised in the book and some not, which readers might profitably mull over as they digest the contents. Are probabilities the most appropriate way of representing the uncertainties which plague decision-making? What is the appropriate role of analysis – to replace judgement, to support judgement, or to model judgement? What are the credits and debits when decision analysis is compared with cost benefit analysis, or with the *soft* problem structuring methods of OR, as a means of making *messes* tractable?

There are other issues aplenty. The subject is alive and kicking. This book provides an excellent way in.

Jonathan Rosenhead

1
Introduction

1.1 PREAMBLE

If I had to pick the technique or methodology that had been most misunderstood by the operational research profession, then I think that decision analysis would be my choice. Certainly it would be if I confined my attention to the operational research profession in the UK. How often have I heard decision analysis berated because it supposedly applies simplistic ideas to complex problems, usurping decision makers and prescribing choice!

Yet I believe that it does nothing of the sort. I believe that decision analysis is a very delicate, subtle tool that helps decision makers explore and come to understand their beliefs and preferences in the context of the particular problem that faces them. Moreover, the language and formalism of decision analysis facilitates communication between the decision makers. Through their greater understanding of the problem and of each other's view of the problem, the decision makers are able to make a better informed choice. There is no prescription: only the provision of a framework in which to think and communicate. Decision makers are not usurped. Far from it: their position is strengthened; they are supported in their task.

It is this view of decision analysis that I wish to convey in the following pages; but, of course, before I can do that I must give you some idea of what concepts it uses and of what techniques lie at its heart. Only when these are before you, can I begin to show you how they combine to help and support decision makers. So let me introduce you to decision analysis – or, rather, let me call upon P. G. Moore and H. Thomas to perform the introduction. Their 1973 paper, The Rev Counter Decision, is quite simply the best introduction to decision analysis that I know. However, perhaps they will forgive me if I emphasize that it is an introduction. You will see a little, but only a little of the subtle, delicate exploration of beliefs and preferences that I alluded to above. In order to present the central ideas in an easily digestible form Moore and Thomas simplify, and thereby risk portraying decision analysis as precisely the naive and simplistic methodolgy that I, and they, deny it to be. When reading their paper remember that you are only seeing an initial outline sketch: there are plenty of hues and tints to be applied before the picture of decision analysis will be complete.

References are cited in these notes and readings in two distinct ways. Those occurring in the text that I have written are indicated by author and year of publication. They are listed in alphabetic order in the references section at the end of the book. Those occurring in the papers reprinted here are indicated by a superscript and listed at the end of each paper.

1.2   The Rev Counter Decision

P. G. MOORE and H. THOMAS
London Business School

This case-based article arose out of a decision tree analysis carried out by the authors in an industrial firm. The company's identity and the actual product manufactured have both been altered for reasons of confidentiality. The various discussions outlined in the case are reported in the same order as they occurred within the firm whilst the problem was originally under study. The work reported was first carried out in 1971 and has been the subject of regular "follow-up" analyses since that date. A much simplified and restricted version of Parts I and II of this case was originally published in two articles appearing in the *Financial Times* during January 1972.

PART I

It is 10.30 a.m. and a meeting is in progress in a Midlands factory office. The Pethow car component company believes there is going to be increased demand for one of its products, a revolution counter for cars. The managing director and four of his executives are considering ways of coping with this new level of demand. Existing plant in the company is working at full capacity on normal shifts, and the firm is considering two alternative options to meet the demand. The first is to expand capacity by putting all its employees on overtime, whilst the second is to purchase an additional revolution counter assembly machine. The managing director has ruled out subcontracting the work to another supplier, because this might result in the subcontractor marketing his own competitive instrument with similar technical features. Furthermore, a price change, except one linked directly to inflation, is ruled out by the marketing manager because of various undertakings that have been given to customers.

Once the options have been outlined, the meeting gets down to a discussion of what might happen under each of them. First of all, they decide that they ought to base their decision upon the gains to Pethow over a 1-year planning period. The marketing manager feels that the projected rise in demand in this period will probably be 15 per cent (if present trends continue), but he adds that there is some possibility of a fall of 5 per cent if the market turns sour. Other possibilities are, he feels, so unlikely that they can be excluded from further consideration. Pressed by the others, the marketing manager admits that he and his staff have accumulated a considerable body of information of future levels of demand and, as a synthesis of this information, he puts the relative likelihoods of the two possible outcomes at something a little different from evens, say 3:2 in favour of increased sales.

Next, the accountant is asked to cost the various options according to the possible outcomes. This he does, after discussions with the production manager on material and equipment costs, and with the personnel manager on wage rates. The new equipment option is costed to include a fair market rent for the use of equipment in the year concerned. The managing director now puts all the data together into a payoff table given below, giving the net cash flows for each possible action and outcome combination, converting the likelihoods to probabilities that sum to unity over all possible outcomes. After a few moments thought he includes a further alternative, namely that they do not accept any order beyond a level that could be met with their present capacity, working no overtime.

Action	Demand (probabilities in parentheses)	
	5 per cent fall (0·4)	15 per cent rise (0·6)
New equipment (S_1)	130	220
Overtime (S_2)	150	210
Existing level (S_3)	150	170

Figures in £000

The managing director asks how they should best handle the information contained in this table, whereupon the production manager suggests that it should be put out in the form of a "decision tree" (Figure 1). In explaining to the others the relevance of the decision tree, he points out that each path along the tree represents a route along which the decision-maker could drive. Along each route he may have to pay a toll, such as the cost of some capital investment, and will attain different outcomes or rewards through following each route. Obviously the tree simplifies even this particular problem but "we can always", the production manager states, "add these complexities later".

The marketing manager next takes up the story, pointing out that the current decision is but one of a whole range of decisions about various products that the company makes, and that their company philosophy is to achieve the highest expected overall gain. This would be satisfied through using Expected Monetary Value (EMV) as a choice criterion at every stage in a project, always choosing that action for which the EMV was highest. To implement the EMV approach they need to look at each possible option in turn. Thus for S_1 (which is to install new equipment) there is a payoff of 220 with probability 0·6, or an alternative payoff of

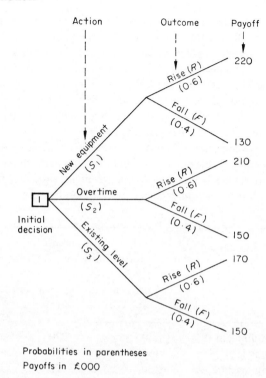

Probabilities in parentheses
Payoffs in £000

Figure 1.

130 with probability 0·4. The EMV would accordingly be:

$$0·6 \times 220 + 0·4 \times 130 = 184.$$

A similar calculation shows that the EMV for the decision S_2 (to have overtime working) is 186. Finally, S_3 (to do nothing, but continue with present production level and no overtime working) has an EMV of 162. The Pethow company ought, under the EMV approach, to choose the action which gives the highest expectation, namely to work overtime. In this instance, of course, the EMVs for overtime and new equipment are so close that the results are virtually indistinguishable. On seeing this result, the personnel manager points out that overtime would be popular with the men, whilst the accountant states that the overtime option would make his life easier as no fresh capital has to be obtained. Hence there seems to be no reason to reverse the strict order to the options produced by the analysis.

The accountant now suggests they should examine the effect that the precise values of the likelihoods assumed have had on the decision reached. To do this, he suggests that the decision should be approached the other way round, calculating the EMV for each of the three options against a range of values of p, the probability assigned to having high demand (the probability of reduced demand would then be $1 - p$). The EMV for the three options would then be:

$$S_1 \quad 90p + 130,$$
$$S_2 \quad 60p + 150,$$
$$S_3 \quad 20p + 150.$$

Thus option S_2 always dominates S_3; whilst S_1 is the best provided that:

$$90p + 130 > 60p + 150$$

or

$$p > \tfrac{2}{3}.$$

If $p < \tfrac{2}{3}$, option S_2 is best. The differences in EMV are very small, around $p = \tfrac{2}{3}$, and hence the suggested decision seems to be relatively insensitive to small changes in the likelihoods assigned to the two possible levels of demand. Accordingly, the meeting is about to break up at 12.45 p.m., having decided to proceed on an overtime basis.

At this stage the managing director intervenes. "Surely", he says, "we have simplified rather too much and glossed over too many of the realities and difficulties. Whilst I can accept that the payoff values calculated are probably correct to within 1 or 2 per cent, and that the decision is not particularly sensitive to the probabilities, I feel that we haven't explored some of the other issues fully enough. For example, shouldn't we have looked at a somewhat longer horizon than one year only; or again, shouldn't we have examined a little more critically whether or not EMV is the appropriate criterion for decision?" The others agree that further examination is necessary and decide to hold the decision over and resume discussion after lunch.

PART II

When the meeting resumes after lunch, the accountant and marketing manager are already hard at work on an extended tree diagram. They explain that they are trying to extend the analysis to a more realistic 2-year planning period, considering only initial options S_1 (new equipment) and S_2 (overtime), arguing that the earlier analysis effectively ruled out option S_3 (do nothing). The marketing manager feels that, in the second year, he needs three levels of possible sales, rather than two levels, to describe accurately the possible situations that could arise. The tree structure that they reach, without any numbers inserted on it, is shown in Figure 2.

The meeting now discusses the structure of the tree. At decision point [1] the choice is between options S_1 and S_2. If S_1 is chosen and sales rise, decision point [2a] is reached, when options S_4 (more new equipment) and S_5 (retain existing equipment but work overtime) are open for choice, similarly along other branches. The general feeling is that, although this tree clearly involves some degree of simplification of the basic problem, and of the options available, the structure provides a reasonable basis for further analysis and lays bare the essential elements of the problem.

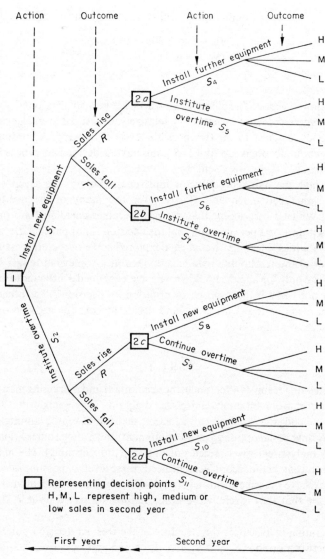

Figure 2.

The group now turns to assessing the numerical quantities required. First, they consider the difficult task of probability assessments. Both the marketing manager and the accountant feel that the rev counter has a good long-term sales future. They argue that, if the sales fell over the first year, the probabilities of high, medium and low sales in the second year are 0·4, 0·4 and 0·2 respectively. If the sales rise in the first year, the corresponding probabilities are 0·5, 0·4 and 0·1 respectively. These probability assessments are "educated guesses" based on the best inform- ation currently available, but nevertheless the managers concerned feel that they provide the best summary they can give of their present views.

The next step is to estimate the various payoffs, or net cash flows that will accrue, if the company "drives" along the many possible routes in the tree. Altogether there are 24 possible payoffs to be estimated, each corresponding to an end position on the right-hand side of the tree, the payoff relating to the whole 2-year period from the start point, marked [1] on Figure 2.

At this stage the managing director comments that the analysis is going to be rather more difficult than for the 1-year problem considered that morning. However, the production manager suggests that the principles used earlier could be extended to the 2-year problem. He argues that if EMV can be used for the single-stage problem, it can equally well be used for the two-stage problem. The essence of his argument is that, when one of the decision points $[2a]$–$[2d]$ is reached the decision criterion will still be EMV. Hence the tree should merely be analysed backwards, from right to left using the EMV criterion for choice, looking at $[2a]$–$[2d]$ first. The production manager says he believes that this backwards analysis procedure is colloquially called "rollback". Decision point $[2a]$ is then considered. The EMV for option S_4 (install further equipment) is:

$$0·5 \times 410 + 0·4 \times 395 + 0·1 \times 380 = 401$$

whilst the EMV for option S_5 (institute overtime) is:

$$0·5 \times 425 + 0·4 \times 408 + 0·1 \times 395 = 415·2.$$

Hence the better decision is to choose option S_5; accordingly a bar is placed on route S_4 and the value of 415·2 placed against decision point $[2a]$, as shown in Figure 3 which reproduces only the upper part of the original tree.

A similar calculation for decision point $[2b]$ shows that option S_7 has a higher EMV (330) than option S_6 (316), and hence that S_7 should be preferred. S_6 is accordingly blocked off and the value of 330 placed against decision point $[2b]$. The two branches leading to the decision points $[2a]$ and $[2b]$ can now be combined. Since the probabilities of a rise or fall in demand in the first year are 0·6 and 0·4 respectively, the overall EMV for the initial option S_1 at decision [1] is:

$$0·6 \times 415·2 + 0·4 \times 330 = 381·1.$$

At this point the managing director asks the other to excuse him whilst he makes a couple of telephone-calls, suggesting that meanwhile they deal with the lower

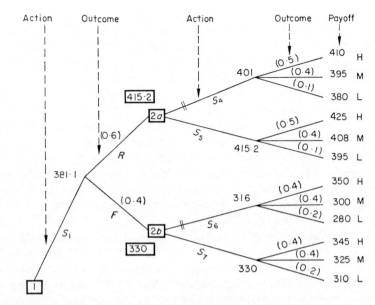

Figure 3.

part of the tree. On his return about fifteen minutes later, they tell him that the EMV for the initial option S_2 (work overtime) is now 340·4 and that Figure 4 gives the outline of the completed tree.

The analysis appears to suggest that the best initial decision now is to install new equipment, a reversal of the morning's decision, and the meeting turns to a discussion as to why this should be so. The reason rapidly emerges, namely that on a 1-year basis overtime is better because the investment cost may not be fully matched with immediate sales increase. In a 2-year period, the growth in sales volume can well make up for any sales shortfall experienced in the first year.

The personnel manager is showing signs of unease at this point. He is unhappy because the meeting seems to be committing the company to massive overtime in one year's time. What if wage rates have soared ahead by then and overtime could only be obtained at a crippling rate? The accountant thinks about this for a moment and replies that at this moment they are only deciding the initial decision, using the best information available to them. If, in 1 year's time, fresh information is to hand concerning the second year's prospects, then they would at that time have to take it into account when making their decision at either point [2a] or [2b]. They are not, he emphasizes, rigidly fixing the later decision now.

The managing director now raises his earlier point of the validity of EMV as a means of deciding between alternatives; "Things always seem to go wrong and I don't fancy working on probabilities" he said, "let's work on the basis that the worst happens." They all peer at the decision tree and, after a moment's pause, the

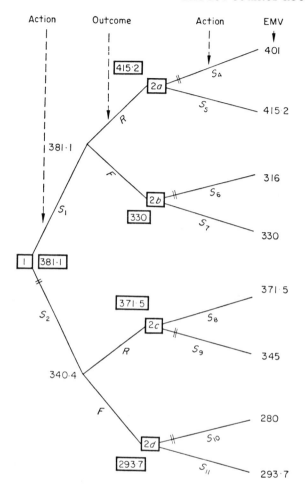

Figure 4.

marketing manager points out that if you take the lowest net payoff for *each* of the
eight possible decision routes through the tree these low values range from 280 up
to 395. On a pessimistic basis the route for which this is highest should be chosen,
i.e. the 395 route, since there cannot thereby be a payoff below that figure. Hence
on this basis, the marketing manager argues, the choice of S_1 for the initial decision
is again the best. The managing director feels reassured by this result, but the
accountant points out that, whilst it is reassuring for this particular situation, he
was under the impression that such coincidence of results did not always occur.

The managing director proposes that, as time is pressing, they should accept the
initial decision to install new equipment for this particular situation, but that they

should review their decision procedures for future situations in the light of the day's experience. In particular, he proposes that they should explore later not only the question of criteria, but also the need to consider discounted cash flows rather than net payoffs and, finally, the way in which extra information becoming available, for example further market research results, could be incorporated in a decision. The meeting breaks up at this point with the managing director agreeing to fix a date for the further follow-up session.

PART III

About a month later, in a meeting at Pethow, Mr Pakin from an associated company which has some experience in using EMV is present. Mr Pakin suggests that the only reason why EMV is not being accepted automatically as the appropriate logical criterion by Pethow is that the executives are all unconsciously recognizing that in many circumstances their attitude towards money varies according to the capital already possessed. Thus suppose you are personally forced to choose between either.

(a) a loss of £10,000 with a probability of 0·001, or
(b) a loss of £15 with probability of 1, i.e. certainty.

"I guess that most of you would choose (b) because you couldn't afford to stand the loss under (a), no matter how unlikely it is. Indeed, you probably already do just this in paying for fire insurance on your house, where the premium is more than the strict expected cost of a fire (because of office expenses, profit, etc.) but you do this with reasonable willingness to avoid the admittedly remote possibility of a loss of £10,000." For the insurance company the situation is rather different. After paying commission and expenses it is left with the expected cost and its profit (provided it got its sums right). It is happy with this because it has large capital assets, insures a large number of houses and can look forward to adding the appropriate overall expected profit to its assets. Hence the individual (and similarly with a small firm) has an attitude that is conservative towards losses, whilst the insurance company (like many large organizations) is more prepared to operate on an EMV basis. Thus many large firms with a large number of small but separate buildings, offices, etc. may well decide to carry their own insurance rather than pay the excess above expected cost that would be required if it were given over to an insurance company. This line of approach can be summarized by saying that, when considering a proposition where the rewards and costs (or losses) are small in comparison with the total assets, EMV is normally appropriate. Where this is not the case, the approach can still be used if *utility* (or some other risk aversion criterion) is substituted for payoff.

The managing director at this point raises again the question of discounting, saying "it seems to me that we are also ignoring something vital, namely the timing of incoming cash and outgoing expenditures". Mr Pakin accepts the point immediately suggesting that "the vital consideration is to get all items on to a common time basis by using an appropriate discount rate: if you commonly use a 10 per cent rate of interest for your appraisal calculations, then use this for each outcome to bring the cash flow to some common base date". Naturally any such move will only make significant differences when the alternative courses of action have cash flows with very different time profiles. For example, overtime working clearly involves roughly equal additional expenditure over the whole time period concerned. Providing extra equipment involves greater expenditure at the beginning of the year concerned. Hence in analysing Figure 1, we could expect that, if discounting were incorporated, the value of S_1 would fall relative to that of S_2, and hence S_2 would become even more attractive than it was before. Discounting should, therefore, always be incorporated when the time scale of the entire process is more than a few months.

At this stage, the managing director says, "I think we are all no doubt gasping for breath, so perhaps we should have a break for lunch now, pursue Mr. Pakin over lunch to elaborate on the points already raised and return to the outstanding matters afterwards." The marketing manager interjects to clarify one point with Mr Pakin before lunch. "What if we consider a planning period of longer than two years, does this affect the analysis?" Mr Pakin states that the analysis would be exactly the same as before except that the structure of the tree would become more complicated. At this point the managing director suggests that they really must get some lunch.

PART IV

After lunch Mr Pakin and the executives of Pethow resumed their discussion by looking into the problem of incorporating fresh information into a decision analysis. To illustrate the argument Mr Pakin suggested that they should look again at the original problem that Pethow considered and suppose that, after the decision tree had been drawn, and the various quantities inserted, the marketing manager wonders whether or not to call in the services of a consultant who has worked for the firm on a number of occasions in the past. This consultant is an expert in the field of the marketing of car components and he offers for a fee of £5000 to sound out, in depth, through his various contacts the possible outlets and come back with a recommendation (or view) as to whether the market for rev counters will rise or fall in the next year. The consultant is by no means infallible, but he has acted as a go-between on a number of deals of this kind and is known to have a reasonable record in these matters. The consultant will, we assume, provide a report that is either favourable (rise) or unfavourable (fall) and the marketing manager summarizes, after some thought, his views of the consultant's reliability in the manner shown by the following table:

| Market outcome | Consultant's report | | |
	Favourable	Unfavourable	Totals
Rise	0·9	0·1	1·0
Fall	0·2	0·8	1·0

This table is read horizontally: thus, should the true market outcome be a rise, there is a 0·9 chance that the consultant will have reported favourably, and only a 0·1 chance that he will have reported unfavourably. Conversely, if the market is going to fall, there is a 0·2 chance that the consultant reports favourably and a 0·8 chance that he reports unfavourably. Naturally, we would like the consultant's matrix to read:

$$\begin{matrix} 1·0 & 0 \\ 0 & 1·0 \end{matrix}$$

but this is a counsel of perfection that will virtually never be achieved in practice and we must take account of the realities of the situation.

The next step, Mr Pakin argues, is for us to see how using the consultant will affect the original probabilities that the marketing manager has assigned to the possible outcomes. These original probabilities are referred to in the literature as *prior* probabilities. Suppose that the consultant is used and he reports favourably. What is now the chance (referred to usually as the *posterior* probability) that the market will rise?

The possible combinations of market change and consultant reports that can occur are sketched in Figure 5. The four different rectangular areas in the diagram represent the expected proportions with which each type of combination of market/consultant report can occur. In the instance we are now envisaging the consultant's report was favourable. Hence only the unshaded area need be considered as the shaded areas correspond to an unfavourable report. Of the unshaded area, the proportionate part corresponding to a rising market is:

$$\frac{\text{Area EFKD}}{\text{Area EFKD} + \text{area GHCK}} = \frac{0·9 \times 0·6}{0·9 \times 0·6 + 0·2 \times 0·4} = 0·87.$$

In other words, the knowledge that the consultant's report was favourable has lifted the probability of a rising market from a prior probability of 0·6 to a posterior probability of 0·87. A similar calculation shows us that the revised probability of a falling market would be 0·13. These two probabilities add up to 1, as they must. If the consultant's report was unfavourable, then the shaded rather than the unshaded area needs to be considered and a similar form of calculation carried out to obtain the appropriate posterior probabilities. The complete set of

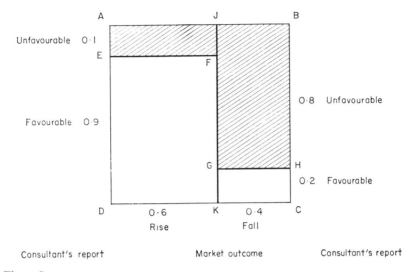

Figure 5.

posterior probabilities is as follows, with the corresponding prior probabilities shown in parentheses.

	Consultant's report	
Market expectation	Favourable	Unfavourable
Rise	0·87·(0·6)	0·16 (0·6)
Fall	0·13 (0·4)	0·84 (0·4)
Totals	1·00 (1·0)	1·00 (1·0)

Inspection of this table shows that a far greater discrimination of the two market possibilities has now been obtained than without the consultant's report, and this should in itself serve to refine the decision-making procedure. The production manager asks whether there is any formalized way by which these posterior probabilities can be calculated, particularly when there are rather more categories to consider. Mr Pakin assures him that the calculations are a straightforward application of a probability theorem known as *Bayes' theorem*.

The accountant speaks up at this point. "I quite see the logic of your approach", he says, "but surely, before we go ahead to see if this changes our original decision, we should bear in mind the cost of the information. This information isn't free, and if it costs too much, it could outweigh any advantage there was in deciding to take one action rather than another." Mr Pakin replies that he did suggest a consultant's fee of £5000 and perhaps they should see how this, in combination with the accuracy of the report, might affect the original 1-year analysis. To understand the revised situation Mr Pakin sketches Figures 6, which is a modification of Figure 1, so as to include two distinct levels of decision: first, as to whether or not to retain the consultant, whilst the second is which of the three possible production options should be taken. The appropriate payoffs, together with the revised probabilities just calculated have been entered.

Some of the tree can be evaluated very simply. The previous analysis corresponded to "no consultant" and led to an EMV of 186 for a production decision to institute overtime. Hence, if the consultant is to be worth his fee, the

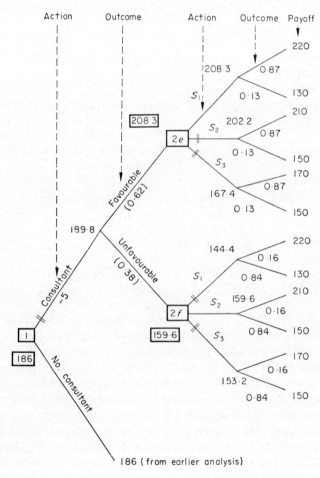

Figure 6.

EMV of the alternative upper branch of the tree must be at least 186, after taking into account the cost of hiring the consultant. Decision points [2e] and [2f] must now be analysed. Both can be treated in the same way as before. Thus, for point [2e], the three appropriate EMVs are:

$$S_1 \quad 0.87 \times 220 + 0.13 \times 130 = 208.3,$$
$$S_2 \quad 0.87 \times 210 + 0.13 \times 150 = 202.2,$$
$$S_3 \quad 0.87 \times 170 + 0.13 \times 150 = 167.4.$$

Hence, S_1 has the highest EMV and both S_2 and S_3 are accordingly barred off on the diagram. Similar calculations made when the consultant's report is unfavourable lead to S_2 (overtime working) having the highest EMV of 159.6. Notice that the two possible consultant reports lead to different actions being recommended – if they did not it would be obvious straightway that the cost of the consultant's services was not worthwhile.

The two best decisions corresponding to points [2e] and [2f] can now be combined, by weighting them according to the likelihood of getting a favourable or an unfavourable result respectively. Consider once again Figure 5. The unshaded portion represents the probability of getting a favourable result from the consultant and this area is $0.9 \times 0.6 + 0.4 \times 0.2$ or 0.62. Similarly, an unfavourable result from the consultant corresponds to the shaded area and is equal to 0.38. These two probabilities add to unity, as they must. Hence the overall EMV for using the consultant is

$$\underset{\substack{\text{Favourable} \\ \text{report}}}{0.62 \times 208.3} + \underset{\substack{\text{Unfavourable} \\ \text{report}}}{0.38 \times 159.6} - \underset{\substack{\text{Consultant's} \\ \text{cost}}}{5} = 184.8.$$

This is below the corresponding EMV when not using the consultant, and hence the consultant appears to be not worthwhile. If he halved his fee, he would be just worthwhile (or if he improved his accuracy he might again be worthwhile). Thus the analysis has provided a way of evaluating the expected worth of the information that can be provided through the consultant.

The managing director felt that the time had come to sum up the day's proceedings, and indeed the whole exercise from which the discussion had emanated. "I believe that we have hit upon a very valuable approach" he said, "which should do two things for us. First, it will encourage us to look at the complete logical structure of any decision that we may have to make. Second, it will force us to quantify as many relevant factors as possible and to include these quantities in the decision process. But it will not replace judgement completely, rather it will make more explicit those areas where judgement is required, and thus help to focus the exercise of judgement more clearly than would otherwise be the case. By so doing, we can hope to reduce both the cost of mistakes, and also their frequency."

POSTSCRIPT

Since the case was first written in early 1972, one of the authors has been engaged in further development work on the problems involved within the Pethow organization. Implementation of the decision analysis work has proceeded and the tree now being examined consists of continuous (rather than discrete) fans of outcomes at the various chance forks of the tree. Further, the analysis is now based on a much longer planning horizon than the 1 year or 2 years described in the case. The overall result of these developments has been to focus attention on two crucial issues in decision analysis and its implementation.

First, and most importantly, has been the need for the measurement of the decision-maker's subjective probability for the various events. Managers have accepted the relevance and efficacy of probability as the language of uncertainty, but have found some difficulty in translating their subjective feelings of uncertainty into probability assessments. Work has been continuing successfully on developing effective operational methods for subjective probability assessment. Secondly, the need has appeared to devise methods to decompose complex trees into simpler and more understandable but relevant forms to stop them becoming a "bushy mess". This applies particularly to longer term sequential decision problems which can rapidly become extremely tedious to handle in a complete form.

REFERENCES

Readers interested in following up the principles of decision analysis that are exemplified in this case are referred to the following brief reference list of books published in recent years. The first three all provide a full development of the basic concepts of decision analysis with a wealth of illustrative material. The last book is of a more concise nature and concentrates on the framework of decision analysis as it might be viewed by a manager.

[1] D. V. Lindley (1970) *Making Decisions.* Wiley, New York.
[2] P. G. Moore (1972) *Risk in Business Decision*, Longman, London.
[3] H. Raiffa (1968) *Decision Analysis.* Addison-Wesley, Reading, Massachusetts.
[4] H. Thomas (1972) *Decision Theory and the Manager.* Pitman, London.

1.3   DECISION ANALYSIS AND OPERATIONAL RESEARCH

We shall discuss, develop and, to a certain extent, justify the theory underlying decision analysis in the coming sections. Before doing so, however, it would advisable to make a few general remarks on its purpose; Moore and Thomas have introduced the basic ideas sufficiently for this to be possible. To structure these remarks, I propose to discuss the relationship between decision analysis and operational research and why the misinterpretations and misconceptions that I referred to in section 1.1 have arisen.

In the early years of operational research, as its sphere of activity changed from the military to those of peacetime, there was much discussion of its aims, its *raison d'être*, and its defining qualities. The first few volumes of *Operational Research Quarterly* clearly show a profession seeking an identity. (A few cynics would suggest that the search still continues!) Most of those working in operational research had been trained in the physical sciences; so it is not surprising that the philosophy of science had a profound effect on their thinking. Operational research was conceived of as the application of the scientific method to problems arising in society, commerce, industry and government. The definition printed at the front of every issue of the Journal Operational Research Society aptly summarizes their conclusion, as of course, it was intended to.

> Operational Research is the application of the methods of science to complex problems arising in the direction and management of large systems of men, machines, materials and money in industry, business, government and defence. The distinctive approach is to develop a scientific model of the system, incorporating measurements of factors such as chance and risk, with which to predict and compare the outcomes of alternative decisions, strategies and controls. The purpose is to help management determine its policy and actions scientifically.

Note the clear intention that the operational research process should be scientific.

A defining quality of the scientific method, especially as it was understood in the late 1940s and early 1950s, is that a scientist must be a dispassionate, disinterested, detached, objective observer of a system. He must not interfere. He must simply observe, record, model and report. True, Heisenberg had suggested that in the final analysis there were limits to a scientist's ability not to interfere, not to affect the system he observes; but his intention must always be to interfere as little as possible. Certainly, his intention should never be to change the system he observes.

Given this it was natural that a separation of duties grew up between the operational research scientist and the manager. The former should observe and model systems; he should use his models to predict the outcomes of possible strategies; and he should give these findings as objectively as possible to the manager. The manager would use this information as a basis upon which to decide on a strategy to implement. Decision making was the province of management; operational research sought only to provide information to the decision-making process, but never to affect the process itself. If the manager asked for a strategy that minimized cost, maximized service or achieved some other objective to be identified, then, of course, an operational research scientist should apply optimization techniques to his models in order to predict which strategies would achieve the desired end. Indeed, this use of optimization techniques has come to be seen as one of the prime roles of operational research. Note, however, it is still a role of providing information. Although optimization points to which strategy to select, it does not decide upon that strategy. The decision is implicit in the definition of the objective to be optimized; and that objective is chosen by management in the light of their values.

It was into this conception of operational research that the methodology of decision analysis had to be fitted. From Moore and Thomas's description it should be clear that decision analysis does affect the decision-making process, and, indeed, is intended to. Decision makers' beliefs are modelled and structured: that was the purpose of attaching probabilities to branches on the decision tree. Note, in particular, the use of Bayes' theorem in Part IV of their paper. This theorem seems to tell the decision makers how to revise their judgements of the likelihood of different outcomes in the light of a consultant's view. There is no question of leaving the decision makers to revise their judgements intuitively. Given their prior probabilities and their assessments of the consultant's relia- bility, their posterior probabilities are defined by the mathematics. Similarly, when we come to introduce utility ideas into the theory in section 2.3, we shall discover that decision analysis also models and structures preferences and value judgements, and in doing so encourages their evolution. Finally, the analysis calculates some expectations and seems to tell the decision makers that they should choose the strategy with the largest expectation. In fact I shall argue that no decision analysis is so dictatorial. Its purpose is to help decision makers understand where the balance of their beliefs and preferences lies and so guide them towards a better informed decision. I accept, however, that this interpretation is not entirely clear from Moore and Thomas's description (which was, as I have said, meant as an introduction to, not a definitive statement of, decision analysis methodology). Certainly, I accept that the descriptions of decision analysis current in the 1950s and early 1960s did not emphasize this interpretation. They portrayed a methodology which said simply that, if a decision maker believed such and such, then logically he must choose so and so.

To an emergent operational research profession, decision analysis was an anathema: it was unscientific; it meddled in decision making. It seemed to tell a decision maker what he should decide. The profession's reaction was, with a few notable exceptions, to reject it; operational research developed without taking decision analysis to its heart. That in itself would not have mattered; decision analysis could have and did develop elsewhere.

Unfortunately, in articulating their rejection, some of those at the forefront of operational research misconceived the purpose of decision analysis. They could have argued simply, as indeed several did, that:

> the function of operational research is simply to provide decision makers with information on which to base decisions – it should not concern itself with the decision process itself' (Adelson and Norman, 1969).

Many, however, argued differently. Decision analysis, they said, first of all models the decision maker's beliefs and preferences, and then uses this model to predict what the decision maker will choose. There are then two possibilities:

1. The model is correct; in which case, since the decision maker is available and since he can choose, why do we need a model? Let him get on with it.
2. The model is incorrect; in which case it would be quite wrong to supplant the decision maker.

(These arguments will be rehearsed in greater detail in section 4.1.)

The essential misconception in their arguments is in the interpretation put on 'model'. They thought of a 'model' only in terms of a descriptive, objective, scientific model. Such models depict, but do not influence decision makers' judgements. Decision analysis requires another type of modelling. The models should be allowed and are intended to influence judgements. They derive not from the scientific method, at least not as it was understood then, but from an area of philosophy lying on the common boundary between that of science and that of ethics. Stafford Beer (1963), following Churchman, identified this boundary and the need to explore it:

> Science deals with what is the case, ethics with what ought to be the case... when science is applied to the processes of decision (as it is explicitly in operations research), and since decisions give expression to value judgement, the area of overlap between science and ethics assumes a new importance.

To understand the full import of decision-analytic models we shall need to explore this boundary. But that, perhaps, is jumping ahead too far. For the present note that, whatever decision-analytic modelling is, its purpose is not simply to describe. Decision-analytic modelling is 'unscientific' because its purpose is more than that of pure description. We shall discuss this further in Chapter 4. Until then, try to avoid imposing any conception of purpose on the theory and case studies to follow: do not allow yourself the same misconceptions as were prevalent in the early years of operational research.

I should perhaps remark before closing this chapter that I have been at pains to point to a misunderstanding so that you do not fall prey to the same. My purpose has not been to provide an explanation of a separate, but related issue: namely, why decision analysis has developed and been implemented faster on the other side of the Atlantic than in the UK. Pearman (1987) analyses several reasons for this: notably the different relationships between academia and industry. But that, as I have indicated, is another matter.

2
The theory

2.1 DECISION TABLES AND DECISION TREES

Decision problems can be represented as either decision tables or decision
trees. To mathematicians the two forms are equivalent: but then to
mathematicians interested in topology a coffee cup is equivalent to a
doughnut; practical men have noted otherwise. It is the same here. For the
purposes of practical decision analysis we shall see that there are enormous
advantages to be gained in using the decision tree representation. The
analysis described in section 1.2 used decision trees without question.
None the less, we should not ignore the decision table format. Discussing it
will make a few salient points, and also serve to emphasize the practical
advantages of decision trees.

The idea underlying a tabular representation of a decision problem is
that the consequence of any action is determined not just by the action
itself but also by a number of external factors. These external factors are
not under the control of the decision maker and, moreover, are unknown
to him at the time he makes his choice. By a **state of nature** or, simply, **state**
we shall mean a particular set of values that these external factors might
assume. If the decision maker knew the state of nature that would actually
hold, the **true state**, he could predict the consequence of his choice with
certainty:

$$\text{action} + \text{state} \longrightarrow \text{consequence}.$$

The difficulty is that the decision maker does not know the true state. We
shall assume, however, that he does know what states are possible. For
simplicity, we shall assume that only a finite number of states are possible:
$s_1, s_2, \ldots, s_n$. Note that these n possibilities form a mutually exclusive
partition of the future: one and only one will hold. We shall similarly
assume that the decision maker has only a finite number of mutually
exclusive alternative actions available: $a_1, a_2, \ldots, a_m$. Although we are
assuming finiteness purely for reasons of presentation, decision analyses in
practice seldom need consider an infinite number of states. Finite decision
models are usually requisite, but we must defer discussing such points until
Chapter 4.

As an aside, I should remark on a problem in drawing readings from
several authors together notation. Here I use a for action, s for state and x

for consequence. Other authors in this collection use a different notation. Moore and Thomas (section 1.2) use s for action or, as they had in mind, strategy. Indeed, in a paper of my own reprinted in Chapter 4 I use a slightly different notation from that here. Other than rewrite all the papers in a common notation, I could think of no easy solution. However, now that you are forewarned, I think it is unlikely that you will fall into any great confusion.

Letting $x_{i,j}$ be the consequence of taking action a_i when s_j is the true state,

$$a_i + s_j \rightarrow x_{i,j}$$

we have a decision table: see Table 2.1.

The symbols $x_{i,j}$ stand for descriptions of the possible consequences. In very simple cases, each $x_{i,j}$ might be a single number, perhaps monetary outcomes as in the three-action, two-state table in the first part of Moore and Thomas's paper (p. 3). Usually, however, there are too many aspects to a consequence for it to be adequately represented by a number. Thus we shall interpret $x_{i,j}$, at least for the present, as a verbal description of the consequence given in sufficient detail for the decision maker to understand what the consequence is. We shall make this rather vague statement more precise in section 2.4. Until then, perhaps an example will serve to illustrate our meaning most clearly.

Suppose that you are going to pick your son up from his university at the end of his term. You are not sure whether he will be bringing all his impedimenta home for the vacation or just enough gear to survive. Unfortunately, you cannot contact him before leaving home. Your problem is that you know that your car is not big enough to take all his belongings. If he is bringing everything home, you need to rent a van. Thus you have a choice between driving off to his university in your car or renting a van and going in that. There are two possible states: he is bringing

Table 2.1 The general form of a decision table

Actions	States of nature			
	s_1	s_2	$\cdots$	s_n
a_1	$x_{1,1}$	$x_{1,2}$	$\cdots$	$x_{1,n}$
a_2	$x_{2,1}$	$x_{2,2}$	$\cdots$	$x_{2,n}$
.	.	.	$\cdots$	.
.	.	.	$\cdots$	.
.	.	.	$\cdots$	.
a_m	$x_{m,1}$	$x_{m,2}$	$\cdots$	$x_{m,n}$

everything home or he is just bringing enough for the vacation. The consequences of your choice might be those described in Table 2.2. The point to note is that, irrespective of whether you feel these consequences are realistic, the entries in the table are descriptions, not numbers.

The problem that faces the decision maker is that he wishes to construct a ranking, an order of merit, of the actions based upon his preferences between their possible consequences and his beliefs about the possible states. Note the use of the word 'construct' here. We shall assume that the decision maker approaches his problem in some confusion. He has preferences between the possible consequences of his choice and he has some feeling for the relative likelihood of the uncertainties facing him; but he is unsure of how to rank the actions in the light of these feelings. He does not know how to achieve a rational balance of the risks facing him. As we develop the theory, we shall allow that he may be unsure of his basic preferences and beliefs; he may be confused by several conflicting objectives and by several interacting uncertainties. For the present, however, we shall assume that, by and large, he is fairly content with his preferences between the consequences and his beliefs about the states. He needs help to sort out his thoughts about the problem, to understand the various ramifications of his choice, and through his increased understanding, to construct a ranking of the actions.

Table 2.2 The decision table for the example

	States of nature	
Action	Your son is bringing everything home	He is only bringing enough for the vacation
Go by car	Comfortable journey; need to persuade son to leave some belongings; delay while he repacks; problems for son during vacation when he needs something left at university.	Everything is fine: comfortable journey; son has planned his packing, so has all he needs for the vacation.
Rent and go by van	Uncomfortable journey; expense of renting van; son has everything he needs for the vacation.	Uncomfortable journey; unnecessary expense of renting van; son has planned his packing, so has all he needs for the vacation.

How might we, as decision analysts, help him do this? First, we might suggest to him some principles of consistency that he might wish to be reflected in his judgements and in his decision making. For instance, we might suggest that, if for any three consequences he prefers x to x' and, in turn, he prefers x' to x'', then he should necessarily prefer x to x''. Also if he is indifferent between x and x' and indifferent between x' and x'', then he should be indifferent between x and x''. If he is indifferent between x and x', but prefers x' to x'', then he should prefer x to x''. Finally, if he prefers x to x' and is indifferent between x' and x'', then he should prefer x to x''. In the terminology of decision theory, we are suggesting here that his preferences and indifferences should be **transitive**. To most people transitivity seems a natural property of rational preferences; i.e. if a set of preferences were intransitive, then most people would be inclined to think those preferences irrational, and with good reason. A decision maker with intransitive preferences can, in principle, be persuaded to part willingly with all his wealth (French, 1986; Lindley, 1985).

It might be that some of the decision maker's actual preferences are intransitive. In such cases, it is assumed that, if this were pointed out to the decision maker during a decision analysis, he would reflect upon his preferences and revise them in the direction of transitivity. We shall discuss this assumed willingness of the decision maker to revise his judgements in Chapter 4.

Similarly, we might suggest that the decision maker's beliefs should be transitive: if for any three states he believes s to be more likely than s' and, in turn, s' to be more likely than s'', then he should necessarily believe s to be more likely than s''. Corresponding conditions should also hold when he believes states to be equally, rather than more, likely. Transitivity seems a natural requirement of rational beliefs. Again we shall assume that the decision maker would be willing to revise particular beliefs, if some were shown to be intransitive.

Assuming that the decision maker accepts that his beliefs and preferences should be transitive, it seems equally reasonable that he would accept that the ranking of the actions which the analysis helps him construct should also be transitive. If at the end of the analysis, he holds a to be a better course of action than a' and, in turn, a' to be better than a'', then he should also hold a to be better than a''.

These transitivity requirements are typical of the principles of consistency which we might commend to the decision maker: there are many others. Unfortunately I have neither the space nor the wish to erect a sufficiently detailed notation to describe and discuss all the principles that we might suggest. In any case I have recently published a book which does precisely that (French, 1986). What I propose to do is refer to the literature for such a discussion and simply present you with its conclusion.

The style and structure of a decision analysis which is appropriate for a decision maker depends upon precisely which principles of consistency he accepts as canons of rationality. The Bayesians have identified a set of principles which have won if not universal, at least substantial support among decision makers. Certainly, when one talks of decision analysis, one tends to mean a Bayesian decision analysis. Moore and Thomas's case study was of a Bayesian analysis and everything that follows in these notes and readings is wholeheartedly Bayesian. The adjective 'Bayesian', by the way, derives from the extensive use the analysis makes of Bayes' theorem, which was introduced in the Moore and Thomas reading and which we shall consider further in the next section.

What the Bayesians have done is show from the particular set of consistency principles which they have adopted that the decision maker's final ranking of the actions should be related to his beliefs about the states and his preferences between consequences through two functions: a **subjective probability distribution** and a **utility function**. These functions have the following three properties:

1. The subjective probability distribution, $P(\cdot)$, represents the decision maker's beliefs in the sense that

$$P(s) > P(s')$$

if and only if, after he has made any revisions in his beliefs necessary to make them consistent, he believes that the state s is more likely to occur than the state s'. Moreover,

$$P(s) = P(s')$$

if and only if he believes the states to be equally likely to occur.
2. The utility function $u(\cdot)$, represents the decision maker's preferences in the sense that

$$u(x) > u(x')$$

if and only if, after he has made any revisions in his preferences necessary to make them consistent, he prefers the consequence x to the consequence x'. Moreover,

$$u(x) = u(x')$$

if and only if he is indifferent between x and x'.
3. The final ranking of the actions is given by their expected utilities, where

$$Eu[a_i] = \sum_{j=1}^{n} u(x_{i,j})P(s_j)$$

is the expected utility of action a_i. Thus to be consistent with his stated beliefs and preferences the decision maker should at the end of the analysis rank a_i above a_k if and only if

$$Eu[a_i] > Eu[a_k],$$

and he should rank them equally if and only if equality holds.

For discussions of the consistency properties assumed by the Bayesians and justifications of these results, see, *inter alia*, Lindley (1985), Raiffa (1968), French (1986), Smith (1987), DeGroot (1970), and Savage (1972). Note that I have listed these references in order of increasing mathematical difficulty.

You will remember that in the analysis of the decision tree in Moore and Thomas's paper the possible options were ranked in order of increasing expected values. The expectations were not of utility, however, but of monetary value: EMV was used as the ranking criterion. In Part III of that paper Mr Pakin argued that this was permissible as an approximation

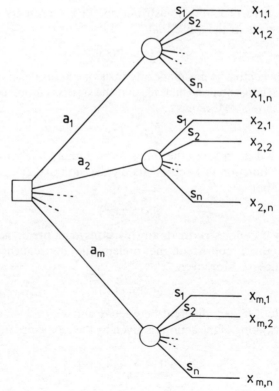

Figure 2.1 A decision tree representation of Table 2.1.

in this case because the rewards and costs associated with the decision were small compared with the company's total assets. We shall discuss the relation between EMV and expected utility further in section 2.3.

The decision table representation of decision problems is important because it is a particularly suitable framework in which to develop decision theory. The consistency properties which should be embodied in good decision making are best discussed within the structure of decision tables. The table format, however, is not well suited to practical analyses. It is static; it pretends or seems to pretend that there is only one point of choice. Yet real-life decisions are dynamic: one decision leads to another, and that to another, etc. In fact, the decision table format is not as static as it first appears. If the decision maker faces a multi-stage problem, the actions in the table are simply defined to be **strategies** or **policies**, i.e. complete contingency plans that prescribe appropriate courses of action as the future unfolds. But that is a theoretical dodge that obscures many issues for practical men: far better to use decision trees.

Decision trees and decision tables are equivalent representations of decision problems (French, 1986). Any decision that can be represented by a table can also be represented by a tree, and vice versa. Figure 2.1 gives the decision tree representation of the general decision problem represented in Table 2.1.

The square on the left of the tree represents the choice between actions $a_1, a_2, \ldots, a_m$; each branch emanating from this square represents an action. The circles are chance points representing the uncertainty about the state of nature, and at these each branch subdivides into n further branches, one for each state. At the end of these are the ultimate consequences, $x_{i,j}$. (In Moore and Thomas's paper chance points were represented by simple points, not circles.)

Given the decision tree format, it is an easy matter to represent multi-stage decision problems by introducing further decision points later in the tree. Consider Figure 2 in Moore and Thomas's paper. The initial decision is represented by point 1, where a choice is made whether to install new equipment or to institute overtime. Possible subsequent choices are represented by decision points 2a, 2b, 2c and 2d. Note that the decision makers will face only one of these four subsequent decisions, depending on which initial decision was made and on which state (sales rise or sales fall) transpires. This particular decision tree is symmetric. The choices at 2a, 2b, 2c and 2d are in a very narrow sense the same choice: whether to buy equipment or to authorize overtime. But, of course, this similarity is more apparent than real. The decision makers face a very different decision at each of these points by virtue of their different situations and knowledge of the market. At 2a they know that they have installed new equipment in the first year, that sales rose, and they know that the financial position is, if not rosy, at least progressing comfortably. At *2d* they know that they introduced overtime in the first year, that sales fell, and that their financial position is not quite so healthy as it would have been at *2a*. Thus they would see the choices at *2a* and *2d* in very different lights, however similar they might at first appear.

In general, decision trees are not symmetric. Decision points in one branch do not necessarily have their counterparts in other branches. Some decisions can only occur if certain previous decisions have been taken. Suppose that I have to travel to London from Manchester. I might have the choice between driving or flying to Heathrow and then travelling into London by tube or coach. I only have to make a choice between the tube and coach if I choose to fly to London: see Figure 2.2. Thus decision point 2 does not have a counterpart in the other main branch of the tree. Decision trees can represent contingent decisions in a very transparent and clear fashion.

We have emphasized a dichotomy between the decision table and decision tree representations. Moreover, we have suggested that in practice decision trees are the more useful. There is a growing school of thought which disagrees with these points on two counts. First, they would argue for a third representation: **influence diagrams**. Second, they would claim that influence diagrams have as much, if not more, to offer as representations for practical analyses, particularly in the case of large problems in which decision trees can fast become bushy messes. We shall not discuss influence diagrams here because, in my opinion, they have yet to prove their worth. Smith (1987) provides an introduction to the main ideas.

Once the decision problem has been represented by a decision tree, the next stage in the analysis is to assess subjective probabilities and utilities. We shall be discussing these assessment processes in the coming sections. For the present it is sufficient to note that probabilities are attached to each branch stemming from each chance point in the tree, as was done in the Rev Counter decision analysis. Utilities should be assessed for each

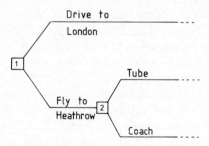

Figure 2.2 Decision tree for the mode of travel example.

consequence, thus replacing the verbal descriptions $x_{i,j}$ at the ends of the final branches with numbers $u(x_{i,j})$. In the Rev Counter decision it was assumed that the utility of a consequence was equal to its projected financial value.

With probabilities and utilities (or financial values) attached to the tree the analysis proceeds in a very straightforward manner. Expectations are simply rolled back through the tree. This procedure of backward analysis is so well illustrated in Part II of Moore and Thomas's paper that it is hardly worth discussing it further here. Many will recognize it as a simple application of dynamic programming ideas. Expectations are calculated at chance points; a branch with the largest expected value is taken at each decision point. The analysis works backward through the tree simply because the later values are needed to calculate the earlier ones. This backward calculation emphasizes a very important point in decision analysis, and in dynamic programming generally; namely, if one decision leads to another, then to analyse the former, one first needs to analyse the latter, since the outcome of the former depends on the choice in the latter. A trivial point it may be, but one that often bears repeating, especially to decision makers so close to a complex problem that they cannot think clearly.

The final stage of a decision analysis should always be to perform a sensitivity analysis. This again is admirably illustrated and emphasized in Part I of Moore and Thomas's paper. True, their sensitivity analysis is not very sophisticated, but it is most certainly present. Alas, the same cannot always be said for other introductions to decision analysis. Sensitivity analysis is often not even mentioned, much less emphasized. Yet it is of central importance, because, as we shall discuss in Chapter 4, it is through a sensitivity analysis that a decision maker gains much of the understanding that a decision analysis brings.

2.2 SUBJECTIVE PROBABILITY AND BAYES' THEOREM

In the previous section we referred to the probabilities that are attached to branches in a decision tree as subjective probabilities. The adjective 'subjective' is important. The probabilities used in a decision analysis are founded upon a quite different conceptual base to 'frequentist' probabilities. In school, college and university the majority of courses which introduce probability theory motivate the concept through notions of relative frequency. In one sense this is quite reasonable: the majority of scientific and statistical theory interprets probability in this way. In another sense it is quite unreasonable: there is a growing body of opinion that any frequency notion of probability fails to be logically and conceptually self-consistent. We shall avoid entering that controversy here, however. The interested reader may find many debates in the literature: e.g. French (1986), Kyburg (1970), and Barnett (1982). For our purposes it will be sufficient to distinguish between the two approaches and to suggest why the subjective one is the more suitable for decision analysis.

The frequency view of probability is that it is the long-run relative frequency with which a system is observed in a particular state in a series of identical experiments. By a series of identical experiments the frequentists mean a series of repeated observations of the system made under essentially similar external or exogenous circumstances. To take a simple example, if a die is thrown many times and the proportion of sixes is seen to tend to 0·1667, then the probability of a six is said to be 0·1667. There are only two points that need concern us about this conception of probability. First, it requires that such an underlying series of repeated experiments is possible. Second, a frequentist probability is an objective property of the system under observation.

To a subjectivist probability represents an observer's degree of belief that a system will adopt a particular state. There is no presumption of an underlying series of experiments. The observer need only be going to observe the system on one occasion. Moreover, a subjective probability encodes something about the observer of the system, not the system itself. We shall discuss shortly how subjective probabilities encode an observer's degrees of belief and, indeed, what is meant by a degree of belief.

Given its emphasis on a series of repeated experiments, it should be clear that the frequentist concept is quite unsuitable for modelling the uncertainty present in the majority of decision problems. Consider the Rev Counter decision. The uncertainty that concerns the Pethow company is whether the demand for their rev counter will rise or fall over a one-year planning period. It is conceptual nonsense to try to imagine the company facing up to a sequence of one-year planning periods while all the exogenous economic, technological and social factors that affect their market remain constant. Hence, it is quite impossible to define frequentist probabilities of the demand rising or falling. On the other hand, subjective probabilities are tailor-made for the purpose; they encode the directors' judgements about the relative likelihoods of the two possibilities. Similarly, the vast majority of decision problems involve uncertainty about events that cannot be embedded in a repeated series of experiments.

We might note in passing that given the emphasis on objective, scientific modelling in the early years of operational research we have here another reason why decision analysis found little favour until recently. It eschews objective, frequentist probabilities and relies on subjective probabilities instead.

The justification for using subjective probabilities in decision analysis does not rest simply on the case that frequentist probabilities are inappropriate. Far from it: the need for something other than frequentist probabilities can be quoted in support of, but not as a justification for, the use of subjective ones. I began our discussion in this section in the way that I did for precisely the same reason that I wrote section 1.3. The ideas in the following pages need to be met with open, fresh minds. Yet many of the terms that I shall need to use will be familiar to you in other contexts where they have other meanings. I need to guard against you imposing those meanings on the ideas here.

The justification for using subjective probabilities in decision analysis rests entirely on the principles of consistency that the Bayesians have suggested should be embodied in rational decision making. In the previous section we suggested that a decision maker faced with a problem should have a feeling of relative likelihood between the states such that he can say whether one state is more likely, equally likely, or less likely than another. We further suggested that the decision maker's feeling of relative likelihood should be transitive. This insistence on transitivity together with some other consistency properties implied that the decision maker's beliefs could be modelled by numbers $P(s_1)$, $P(s_2)$, etc. such that

$$P(s) > P(s')$$

if and only if he believed s to be more likely to transpire than s'. These numbers were called subjective probabilities. It happens that the consistency properties imply that these subjective probabilities behave mathematically like any other type of probability. They obey Kolmogorov's laws: they form a non-negative, additive measure over the field of events with total mass 1; or, less mathematically, the probability of an event is a number between 0 and 1, and the probability of the union of two disjoint events is the sum of their probabilities (DeGroot, 1970; French, 1986; Lindley, 1985; Raiffa, 1968; Savage, 1972).

Subjective probabilities have become known as **degrees of belief** in many parts of the literature. Suppose that for two states s and s' the decision maker believes s to be more likely than s'. It follows from the representation of his judgements given by subjective probabilities that

$$P(s) > P(s')$$

Thus it is a very natural figure of speech to say that he has a higher degree of belief in s than in s', and to refer to $P(s)$ as his degree of belief in s.

It is all very well to suggest that subjective probabilities must exist if the decision maker's beliefs are consistent in a certain sense; but to use subjective probabilities in a decision analysis we need to know more than of their existence. We need to be able to assess them. How may this be done? Moore and Thomas suggested that the particular probabilities used in their analysis were

'..."educated guesses" based on the best information currently available'.

Well, perhaps: but the assessment of subjective probabilities is much more structured, informative and reflective than this would suggest.

Suppose that I ask you to consider the likelihood of rain tomorrow. By 'rain tomorrow' I mean some well-defined event: perhaps that at least 0·1 inch of precipitation will fall in the garden of your home. I do not mind whether you have heard a weather forecast or not; not whether you keep a piece of seaweed or a barometer; nor, indeed, whether you are an amateur meteorologist who has kept records of the weather conditions in your locality for several years. All that matters to me is that you understand what is meant by the event 'rain tomorrow'. Suppose further that I offer you a choice between two bets:

Bet A: £100   if it rains tomorrow;
    £0   otherwise.
Bet B: £100   if a die which you consider fair lands six on the next toss;
    £0   otherwise.

If you prefer Bet A to Bet B, it seems reasonable to suggest that you judge it to be more likely to rain tomorrow than a fair die to land six. Moreover, since you consider the die fair, you should consider all six faces equally likely. Thus your probability that the die lands six should be 1/6, and your probability of rain tomorrow should be greater than 1/6. Similarly, if you prefer Bet B to Bet A, your probability of rain tomorrow should be less than 1/6.

By offering you similar choices between Bet A and other bets based upon the fall of a die, e.g. £100 if an even face shows, nothing otherwise, I can locate permissible values for your probability of rain tomorrow more closely. Essentially, I am asking you to compare the likelihood in your opinion of rain with the likelihood of events based upon the fall of a die.

Discussing throws of a die provides useful motivation for the methods of assessing subjective probabilities, but it does not lead directly to a very practical method of assessment. The problem is that a die which is fair in your opinion generates six equally likely events. Therefore, comparison of Bet A with bets based upon a throw of that die can only locate your subjective probability of rain tomorrow to within 1/6. True, by considering throws of two fair dice I can locate your subjective probability to within 1/36; and throws of three dice allow me to locate it to within 1/216. But you are unlikely to have a feeling for the uncertainties involved in throwing several dice. Snakes and ladders and other simple board games may have given you a feeling for the relative likelihood of events based upon a throw of a single die; *Monopoly* and more complex board games may have provided you with a similar feeling for throws of two dice; but, apart from poker dice, few games will have given you any experience of throwing more than two dice. Thus the assessment of subjective probabilities is usually based upon some other randomizing device than throwing dice.

In French (1986), I use the idea of a **probability wheel** or a **wheel of fortune**, as it is sometimes known. This is simply a well-balanced pointer which may spin on the face of a disc (see Figure 2.3). The disc is such that sectors of various sizes may be indicated on it. It is possible to become familiar very quickly with the uncertainties involved in spinning the pointer and seeing into which sector it ends pointing. It is a more natural randomizing device than throwing dice.

Now, if you accept that the pointer is well balanced, you will surely accept that, if I spin the pointer, it is more likely to end pointing into a sector subtending a larger angle at the centre than one subtending a smaller angle. More precisely, given the choice between the two bets below based upon the probability wheel as shown in Figure 2.3, you should prefer Bet C to Bet D.

Bet C: £100   if the pointer ends in sector C;
       £0   otherwise.
Bet D: £100   if the pointer ends in sector D;
       £0   otherwise.

To assess your probability of rain tomorrow, I would simply ask you to compare Bet A with bets like Bets C and D based upon the spin of a probability wheel. Suppose that for the particular sectors shown in Figure 2.3 you prefer Bet C to Bet A and, in turn, prefer Bet A to Bet D. Then it is reasonable to infer that I can find a sector intermediate in size between sectors C and D such that, if I offered you a bet based upon that sector, you would be indifferent between that bet and Bet A. If that sector subtends an angle of z degrees at the centre, it follows that your probability of rain is $z/360$. There are a number of important points that I should make here.

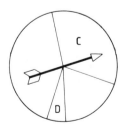

Figure 2.3 A probability wheel.

First, I am well aware that it is unlikely that you could identify a unique value of z which alone leads you to be indifferent between the bets. In principle, I believe – and I hope that you will accept – that you should be able to. If there are two sectors with distinct values of the angle subtended, then you should think the sector with the larger angle more likely than the smaller. Thus you should not think them both as equally likely as rain tomorrow. However, human discrimination is finite and it is unlikely that you would be able to do more than locate a range of values of z for which you are indifferent. Any good decision analysis recognizes this and through a sensitivity analysis checks to see whether the decision maker's lack of precision in specifying probabilities has any significant implications. For instance, in Part I of the Rev Counter study it was shown that provided that the decision makers assessed the probability of a high demand as either clearly greater than 2/3 or clearly less than 2/3, the choice of option was clear. Only if the probability were about 2/3 would there be a difficulty; and that difficulty would be more apparent than real. For in that case it would not make much difference to the company's EMV whether they installed new equipment or instituted overtime.

Second, I have claimed that the purpose of a decision analysis is to provide decision makers with a framework in which to think and so to help them understand their problem, their preferences and their beliefs better. Indeed, I have further suggested that the analysis may lead them to recognize inconsistent judgements and revise these in the direction of consistency. We can begin to see here how the assessment of subjective probabilities can do this in the case of the decision makers' beliefs.

Above I asked you to consider your belief in the event 'rain tomorrow'. Let me complicate things a little. Suppose that you are considering whether to go to a football match tomorrow. Your local team are never exciting to watch in the rain: their play becomes scrappy. So if it rains you definitely do not want to go. To complicate matters, their only player of any worth is injured and it will not be clear whether he will be fit to play until immediately before the match. Without him the team are a shambles and definitely not worth watching. In short, the game will only be worth watching if it does not rain and if the star player is fit. That is quite a complex event to think about: it involves two uncertainties simultaneously.

It is reasonable to suggest that you would agree that your beliefs in the two events are independent. By **independent** I mean that if you were given a perfect weather forecast, one that is guaranteed to be 100 per cent accurate, then this information would not lead you to revise your beliefs about the player's fitness in any way. Similarly, knowing for certain whether the player was going to be fit would not affect your beliefs about the weather. By asking you to agree with this independence, I am not suggesting that you have sorted out your beliefs about the weather or about the player's fitness yet. I am simply asking you to agree in principle that had you done so then they would have been independent in this sense. Now independence in subjective probability theory corresponds precisely with the concept of **probabilistic independence**, which you will have met in other fields of probability modelling. So, whatever numerical values your probabilities eventually take, they must be related by:

P(no rain tomorrow and the star player is fit)

$= P$(no rain tomorrow) $\times P$(the star player is fit)

$= [1 - P$(rain tomorrow)$] \times P$(the star player is fit).

Therefore, to assess your probability of it not raining tomorrow and the player being fit, I may assess your probabilities of two simpler events. The framework provided by the analysis enables us to separate complex uncertainties into simpler ones. Now I may begin to assess your probability for rain tomorrow exactly as I described above. Note how this will allow you to focus on your judgements about the weather without being confused by any other aspects of your problem. In doing so it is likely that you will better appreciate your judgement of the likelihood of rain and your reasons for making it. Similarly, I may assess your judgement about the probability of the player being fit, and again the assessment will allow you to focus on that issue unclouded by any others.

A second example, this time using conditioning, will illustrate how the assessment process can help the decision maker organize his thoughts. Suppose that a research and development manager must decide whether to sanction development of a new product line. The electronics involved could be produced by conventional methods, but new surface-mounting technology might be suitable and, if it were, would produce a revolutionary product. In considering the likelihood of ultimately producing a successful product, the manager must weigh up the chances of the product being successful in the case in which it proves possible to use surface mounting and in the case in which it does not, with the chances of whether surface mounting will be suitable. Put in probabilistic terms:

P(successful product)

$= P$(product successful$|$surface mounting suitable)

$\quad \times P$(surface mounting suitable)

$+ P$(product successful$|$surface mounting unsuitable)

$\quad \times P$(surface mounting unsuitable).

Here P(product successful$|$surface mounting suitable) means the manager's degree of belief in the success of the product in the case that it does prove possible to use surface mounting.

In principle, it is possible for the manager to assess P(successful product) directly. But it clearly helps him organize and focus his thinking if he assesses the simpler probabilities on the right-hand side of the equation and then combines these by multiplication and addition.

There are many other ways in which judgements of likelihood can be **decomposed** into simpler components. The idea being that, although one has to make more judgements, it is easier and more 'accurate', in some sense, to make several simple judgements and combine them by the laws of probability than to make one complex judgement holistically. A recent paper by Ravinder *et al.* (1988) analyses some of the ways in which decomposition may bring greater reliability of assessment. However, to discuss accuracy and reliability is to miss much of the point of decomposition. Breaking down a complex uncertainty into its component parts actually helps a decision maker think. It draws and focuses his attention on each of the component uncertainties before him.

The assessments that we have discussed so far have concerned events in a dichotomy of the future, e.g. it will rain or it will not. Decision problems usually involve a much finer partition of the future. In the case of the weather, it might be necessary to consider the partition {*sleet, snow, rain, cloudy but dry, sunny*}. The addition law of probability gives:

$$P(sleet \text{ or } snow \text{ or } rain) = P(sleet) + P(snow) + P(rain)$$

Such relations make it possible to check the consistency of assessments. If I were assessing your probabilities for the weather tomorrow, I would ask you to consider bets which would enable me to assess all four probabilities appearing in this equation. I would then check that the equation did in fact hold. If it did, well and good; but, if it did not, then I would suggest to you that your beliefs were inconsistent. As I have remarked, it is a tenet of decision analysis that, if you are made aware of such inconsistency, you will reconsider your beliefs and revise them so that they become consistent. It is through processes such as this that decision analysis helps your beliefs evolve. Moreover, it is impossible, I would argue, to revise your beliefs without reflecting upon them and the reasons for which you hold them. Again the analysis is helping you towards a better understanding of yourself.

The third point to note about the assessment procedure concerns the consistency properties demanded of a rational decision maker by the Bayesian school. I am striving to avoid introducing and discussing these properties in full because of the detail that would be required. It seems only fair, however, to remark that one of the most controversial of these underpins the assessment of subjective probabilities. Essentially, the Bayesians suggest that a rational decision maker should be prepared to compare his uncertainties about the true state of nature in his real decision problem with those generated by a randomizing device, such as a probability wheel. In particular, they assume that the decision maker is prepared to consider his choice between bets such as Bets A, B, C and D above. Why should he? They are hypothetical options that he cannot really accept. Why should he waste his time considering what cannot be instead of the options in his true decision problem? The answer to these questions, I believe, depends critically on one's view of decision analysis.

If you accept, as I do, that much of the purpose of a decision analysis is to construct a framework in which to think, then it is reasonable to allow the analysis to bring into the discussion girders, nuts and bolts with which to build that framework. The randomizing device and the bets constructed upon it are vital girders. They allow the analysis to separate some aspects of the decision problem from the rest so that the decision maker can focus his attention upon them. The foregoing examples suggest how this happens in the assessment of subjective probabilities. We shall see in the following sections how a similar use of randomizing devices is used to assess utility functions and in the process helps the decision maker to focus upon aspects of his preferences. Once decision makers have experienced a decision analysis, they come to appreciate the help and understanding that it brings. The case studies that we shall meet amply illustrate this. This appreciation provides the necessary motivation to consider hypothetical bets seriously.

The assessment procedure sketched here underlies most of those used in practice, but it is, to say the least, archetypal. Practical procedures are much more subtle and are designed to avoid problems of anchoring and bias, which we have yet to mention. None the less, I hope that I have conveyed enough of the 'flavour' of the methods by which decision analysis seeks to encode decision makers' uncertainties about the true state to enable you to appreciate the case studies and arguments that follow. Surveys of methods and protocols for assessing subjective probabilities may be found elsewhere in the literature (Hampton et al., 1973; Merkhofer, 1987; Savage, 1971; Spetzler and Stael von Holstein, 1975; von Winterfeldt and Edwards, 1986; and Wallsten and Budescu, 1983).

Anticipating arguments to be presented in Chapter 4, there is much evidence from psychological studies that decision makers without the support of a decision analysis often express judgements which are inconsistent in the Bayesian sense (Hogarth, 1980; Kahneman et al., 1982; von Winterfeldt and Edwards, 1986; Wallsten and Budescu, 1983; Wright, 1984). Assessment procedures ask decision makers to express judgements, albeit easier, more structured judgements than are required of them in their actual decision problem. It is to be expected, therefore, that assessment procedures may encounter difficulties because of inconsistencies in the decision makers' responses. Many of these inconsistencies can be detected and used to advantage by referring them to the decision makers for revision. As I suggested above, by encouraging them to reflect on their judgements in this way a decision analysis helps decision makers understand their beliefs better. Some inconsistencies, however, are harder to detect and assessment procedures are designed to avoid their occurrence. For instance, there is an anchoring effect in which decision makers tend to pull back to an initial assessment when an inconsistency is brought to their attention. Equally, if the analyst should suggest a value to a decision maker, that value will act as an anchor on later assessments, introducing a bias. Further details are given in the above references.

Beliefs are not static; they change as observations are made and evidence accumulated. For the remainder of this section we shall not question that beliefs change, but rather consider how they should change. How should a decision maker revise his beliefs when he learns a relevant piece of information?

First, note an assumption implicit in this question. A decision maker's beliefs are always conditional on his current knowledge. We have written $P(s)$ for a decision maker's subjective probability for a state, s. We have emphasized that $P(s)$ is subjective; it depends on the decision maker. It is equally clear that it depends on s, too. What we have not taken explicit note of is that $P(s)$ also depends on the decision maker's knowledge at the time the probability is assessed. In assessing your probability of rain tomorrow, you will agree surely that your beliefs and, hence, $P(rain)$ would depend on whether you had heard a weather forecast or not. Perhaps we should write $P(rain|H)$, where H is your knowledge at the time of assessment, to emphasize this. Some authors, indeed, do use such a notation; but we shall not, instead relying on our remarks here to make the point sufficiently.

If it were the case that at all points in a decision tree the probabilities represented the decision makers' beliefs conditional on the same knowledge, then we not need to make the above point at all. But that is not the case. The very structure of a decision tree means that the decision maker's information differs between the choice points. Consider the tree shown in Figure 6 of the Rev Counter case study. The directors' information at points 2e and 2f would be quite different. Moreover, the difference would be relevant to their beliefs at later chance points in the tree. At 2e the directors would have a favourable report from the consultant before them. Unless they had wasted their money on a consultant who invariably provides a report full of irrelevant waffle, this surely must change their beliefs about the demand; we might expect that they would accord a high demand a greater likelihood than before they had received the information. Similarly, at 2f they would have an unfavourable report before them, and we might expect their probability of a higher demand to be smaller than before.

It might be thought that this matter need not detain us long. All we need do is make the decision makers aware of the information which they have – or, rather, which they are imagined to have – at each chance point in the tree. Then, when their probabilities are assessed, they must make their judgements in the light of the information available to them. Thus, in assessing their probabilities at chance points after $2e$, the analyst would ask the directors to imagine that they had received a favourable report; whereas at chance points after $2f$ he would ask them to imagine that they had received an unfavourable report. Unfortunately, there is much psychological evidence that decision makers perform very poorly in such assessments. When asked to assimilate a specific piece of evidence, such as the consultant's report, into their beliefs, decision makers resort to very poor heuristics to revise their beliefs. Often they forget all the information that they had prior to the piece of evidence and focus on that entirely (see Hogarth, 1980; Kahneman et al., 1982; and Wright, 1984). In short, decision makers need help to structure the revision of their beliefs; and that is where Bayes' theorem comes in.

You have already have seen Bayes' theorem in operation in the Rev Counter study, although the mechanics of it were disguised in the geometry of Figure 5 of that paper. Let me remind you of the algebra of the theorem and show you how it relates to that geometry.

Suppose that D is a specific event or, if you prefer, a piece of data, the knowledge of which would affect the decision maker's beliefs about the true state of nature in his decision problem. Let $P(s_1), P(s_2), \ldots, P(s_n)$ be his **prior probabilities** for the states before he knows whether D has occurred or not. Let $P(D|s_j)$ be his conditional probability for the occurrence of D given that s_j is the true state; i.e. if he knew that s_j were the true state, then his belief in the occurrence of D would be represented by $P(D|s_j)$. Then Bayes' theorem states that for any particular s_j his **posterior probability** of s_j given D, is

$$P(s_j|D) = \frac{P(D|s_j) \cdot P(s_j)}{P(D|s_1) \cdot P(s_1) + P(D|s_2) \cdot P(s_2) + \cdots + P(D|s_n) \cdot P(s_n)}$$

$P(s_j|D)$ represents his belief that s_j is the true state if he were to learn that D had occurred.

Now look at the calculations immediately following Figure 5 in the Rev Counter study. For the case being considered there, D is the event that the consultant reports favourably. There are two possible states in the problem ($n = 2$): s_1, the market rises; s_2, the market falls. So substituting in Bayes' theorem gives:

P(market rise|favourable report)

$$= \frac{P(\text{favourable report}|\text{market rise}) \cdot P(\text{market rise})}{\begin{array}{l}P(\text{favourable report}|\text{market rise}) \cdot P(\text{market rise}) + \\ P(\text{favourable report}|\text{market fall}) \cdot P(\text{market fall})\end{array}}$$

$$= \frac{0.9 \times 0.6}{0.9 \times 0.6 + 0.2 \times 0.4} = 0.87$$

This is precisely the same numerical calculation as that under Figure 5 of the Rev Counter study, as, of course, it should be; the geometric argument used by Moore and Thomas is no more than a 'pictorial proof' of Bayes' theorem. I leave it as an exercise to show that using Bayes' theorem for the case of an unfavourable report:

$$P(\text{market rise}|\text{unfavourable report}) = 0.16$$

It might be thought that, if decision makers need help to structure their beliefs so that conditional probabilities such as P(market rise|favourable report) can be assessed, they would need similar help before conditional probabilities such as P(favourable report|market rise) can be assessed. Thus Bayes' theorem would seem to translate one difficult problem into several equally difficult problems. In practice, however, this is not so. If the directors were to assess P(market rise|favourable report) directly, they would have to combine intuitively their own views on market prospects with those expressed by the consultant. It is the process of combining information that seems, empirically, to be so difficult. In assessing P(favourable report|market rise), the directors do not have to keep their own views on a market rise in mind. Furthermore, as the case study suggests, the company may have records of the consultant's previous performance and these can be used in the assessment.

Note that the records cannot give information on P(market rise|favourable report) directly. The records provide a history of the consultant's forecasting ability in many different market conditions; not just conditions similar to those prevailing at the time of the decision. Indeed, one way of looking at Bayes' theorem is that the use of P(market rise) and P(market fall) in the numerator and denominator adjusts information obtainable from historical records so that it is appropriate to current circumstances.

As an aside and at the risk of complicating issues, perhaps I should admit that the use of Bayes' theorem to assimilate the information in a consultant's report is not quite as straightforward as the above suggests (see French, 1985). In general, there is no conceptual difficulty in applying Bayes' theorem to guide the revision of beliefs when a new piece of data is received. When that piece of data is the opinion of another person, however, some subtlety is required. But we shall not dwell on that issue here.

2.3   UTILITY THEORY

In section 2.1 we noted that subjective probabilities were only one component of the Bayesian model of rational decision making. They represent a decision maker's beliefs; but it is also necessary to represent his preferences through a utility function, which is a function on the set of consequences such that

$$u(x) > u(x')$$

if and only if he prefers the consequence x to the consequence x'; and, moreover, such that

$$Eu[a_i] > Eu[a_k]$$

if and only if he prefers action a_i to action a_k, where

$$Eu[a_i] = \sum_{j=1}^{n} u(x_{i,j}) \cdot P(s_j)$$

the expected utility of action a_i.

It is important to realize that $u(\cdot)$ is not just any function whose value increases with the decision maker's increasing preferences for the consequences. The function $u(\cdot)$ must be assessed in such a way that the expected utilities of the actions rank the actions in a manner that is consistent with the decision maker's preferences between the consequences and his beliefs about the true state. Since his decision problem involves uncertainty, it is not surprising that the appropriate manner of assessing $u(\cdot)$ involves questioning him about his preferences between carefully constructed bets, i.e. questioning him about his preferences between simple uncertain prospects.

Suppose that x_{best} is the best possible consequence within the decision table representing his problem. Thus he prefers x_{best} to $x_{i,j}$ for any i, j. Similarly, suppose that x_{worst} is the worst possible consequence: he prefers $x_{i,j}$ to x_{worst} for any i, j. Now the origin and unit of a utility function are arbitrary (French, 1986). So we may set:

$$u(x_{best}) = 1$$

and

$$u(x_{worst}) = 0$$

To assess $u(x_{i,j})$ for any other consequence, we offer the decision maker the choice between having $x_{i,j}$ for certain or taking part in a bet based, say, on a probability wheel. For definiteness, imagine that we offer him the following bet based upon the wheel shown in Figure 2.3.

Bet A: x_{best} if the pointer ends in sector C;
x_{worst} otherwise.

Suppose that he prefers Bet A to $x_{i,j}$ for certain. The we might offer him a second bet based on sector D:

Bet B: x_{best} if the pointer ends in sector D;
x_{worst} otherwise.

Suppose now that he prefers $x_{i,j}$ for certain to Bet B. Then it is reasonable to infer that we can find a sector intermediate in size between sectors C and D such that he is indifferent between $x_{i,j}$ for certain and a similar bet based upon that sector. Finally, suppose that this intermediate sector subtends an angle at the centre of z degrees. Then, since the utility function represents his preferences between uncertain prospects through expected utilities and since his probability of winning x_{best} is $(z/360)$:

$$u(x_{i,j}) = (z/360)\cdot u(x_{best}) + (1 - z/360)\cdot u(x_{worst})$$
$$= (z/360)\cdot 1 + (1 - z/360)\cdot 0$$
$$= (z/360)$$

Note how the expected utility property is central to this assessment method. It is because of this that the expected utilities of the actions in his decision problem provide a ranking that is consistent with his beliefs about the states and preferences between the consequences.

Since $(z/360)$ is the decision maker's probability of winning x_{best} in the bet, we are in effect asking him: in a gamble between x_{best} and x_{worst}, what probability of winning x_{best} makes you indifferent between the gamble and having $x_{i,j}$ for certain? In the following extract (Adelson, 1965, pp. 31–35), this form of questioning is used to assess utilities of monetary consequences. In his example, $x_{best} = £1000$ and $x_{worst} = -£1000$.

Utility theory

Modern utility theory is a quite different concept from the utility theory of the classical economists. Whereas the latter attempted to attach a "value" to the possession of certain goods, the modern theory (generally attributed to von Neumann and Morgenstern[12]) is meant to be a portrayal of an individual's attitude to risk. Consider the simplest gambling situation. A ticket in a certain lottery, in which there is only one prize, costs (say) £5. If the prize is £10, how large would have to be the probability of your winning before you would buy a ticket? Suppose the prize were £50 or £100, what would the probabilities be then? Now multiply all the above numbers by 10,000 (or 100,000). Assume the lottery ticket is the cost of the investment in a plant which will either be a complete flop or produce the stated returns as a "prize". What would have to be the probabilities before you would recommend that your firm should go ahead with the investment? *Decision theory does not purport to be of any help in making these decisions; the answers will always be particular to the individual making them.* However, once the answers to certain questions of this sort have been given, the individual's "utility function" can be drawn up. When this has been done decision theory can be used to solve much more complex decision problems (ones in which there are many possible decisions and many states of the world) in such a way that *the answers obtained to such problems will be consistent[†] with the answers that the individual gave to the simple decision questions.*

In order to see how a utility function is derived, imagine that you have got yourself into a situation where you may win £1000 (event *A*) or lose £1000 (event *B*), depending on the luck of a draw. If you are unhappy with the situation, you can buy yourself out at a cost of £*X*. If you are happy with the situation, you can consider selling out. Suppose you are told that the chance of event *A* is 0·1, and that of event *B* is 0·9. Would you prefer to accept the situation, or would you want to buy yourself out? If the latter, what is the most you would pay? Suppose after weighing the relatively high chance of losing £1000, you decide you would be prepared to pay about £850 for a release. Now suppose the probability of *B* is reduced to 0·7, and that of *A* increased to 0·3 correspondingly, what price would you pay then? For the sake of argument we will take this to be about £650. If the chances of events *A* and *B* are 50–50, many people to whom a loss of £1000 would be something of a burden would still prefer to buy themselves out of the possibility by payment of a sum, let us say £200. On the other hand, if the probability of a win rises to, say, 0·8, one might feel that this was a chance worth taking. However, it would be expected that if offered, say, £300 cash, you would be prepared to forgo your chances of £1000 gain (or £1000 loss).

We can now calculate the utilities of these fixed sums, *in this gambling situation*, in the following way. The *end points* of the utility scale are arbitrary. Let us *assign* a utility of zero to a loss of £1000 or (as it is convenient to express it) a gain of − £1000. Let us assign a utility of unity to a gain of £1000. The utility of the sum which would have the same value to the decision-maker as a gamble giving 0·9 probability of zero utility and 0·1 probability of unit utility is calculated as $0·9 \times 0 + 0·1 \times 1 = 0·1$. That is, in the example, the utility of − £850 is 0·1. Similarly, the utility of − £650 is 0·3. The utility of − £200 is 0·5, the utility of £300 is 0·8. We can now draw up the implied utility curve (Figure 2) by passing a smooth curve through the plotted points (in practice we would attempt to obtain more points, and then check the points against the curve, when plotted). Notice that this curve

---

[†]"Consistency" here implies behaviour in accordance with certain axioms, i.e. seemingly obvious elementary principles of choice. For example, "preferences are transitive" – that is, if a person is faced with three alternatives *A*, *B* and *C*, and he says he prefers *A* to *B* and *B* to *C*, then he should prefer *A* to *C* if *B* is withdrawn. For a complete discussion see Chapter 2 of Reference 9.

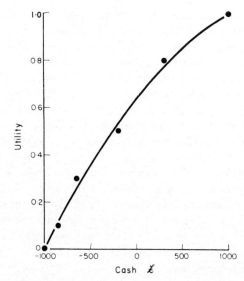

Figure 2 A utility curve.

reflects a conservative attitude towards risk taking, in that a loss of £500 (say) shows a greater fall in utility than the rise due to a gain of £500. Another way of looking at this is to note that the individual concerned would, when faced with a possible risk situation in which he stood to gain or lose £500, require odds of better than 50–50 of a win before he would take the bet. The conservative utility function appears to be virtually universal for "serious" decision-making situations, although other forms are possible.[†] A recent paper[14] which describes an experimental determination of utility curves in an industrial organization brings out this inherent conservatism very clearly.

As an example, consider the following: You have to choose between 3 investment opportunities, each costing £1000 but having different probabilities for the various pay-offs as shown in Table 7.

Utility analysis can now be used to determine which investment should be chosen in order to be "consistent" with the answers given to the questions from which the utility curve was drawn up. The procedure is to replace the cash sums involved by their utilities, as read off from the utility curve. Calculate the expected utility of each investment and choose that showing the highest. Table 8 illustrates the calculations for this example.

[†]In Chapter 4 of Reference 13 it is demonstrated that utility curves generally tend to be sigmoid in shape, if the range of pay-offs considered is sufficiently great.

Table 7. Comparison of investments problem

	Probability of pay-off		
Pay-off	Investment A	Investment B	Investment C
Lose £1000 entirely	0·10	0·20	0
Lose only £500	0·20	0·10	0·35
Break even	0·30	0·20	0·25
Make a profit of £500	0·20	0·20	0·30
Make a profit of £1000	0·20	0·30	0·10

Investment C is the one which should be chosen. Moreover, the decision-maker of this example should be "indifferent" (i.e. would be prepared to choose on the result of a toss of a fair coin) between investments A and B, if C be removed from consideration. The expected cash value of each investment can also be calculated. On this basis, investment C is the worst, having an expected cash value of £75. Investment B is best, having an expected cash value of £150, while investment A has an expected cash value of £100. The fact that utility analysis chooses investment C is simply a reflection of the conservatism shown by the decision-maker when his utility curve was drawn up, i.e. it is preferred partly because it is less risky than either of the other two. Similarly, although the expected cash value of investment B is 50 per cent more than that of investment A, it is also "more risky", as far as possible losses are concerned, and this makes their values, considered as entities, equivalent in the eyes of this decision-maker.

Table 8. Utility analysis solution of problem in Table 7

Cash pay-off (£)	Utility	Investment A		Investment B		Investment C	
		Proba-bility	Weighted utility	Proba-bility	Weighted utility	Proba-bility	Weighted utility
−1000	0	0·10	0	0·20	0	0	0
−500	0·36	0·20	0·072	0·10	0·036	0·35	0·126
0	0·64	0·30	0·192	0·20	0·128	0·25	0·160
500	0·86	0·20	0·172	0·20	0·172	0·30	0·258
1000	1·00	0·20	0·200	0·30	0·300	0·10	0·100
	Expected utility		0·636		0·636		0·644

Table 9. Utility analysis solution of modified problem

Cash pay-off (£)	Utility	Investment A		Investment B		Investment C	
		Proba-bility	Weighted utility	Proba-bility	Weighted utility	Proba-bility	Weighted utility
−100	0·59	0·10	0·059	0·20	0·118	0	0
−50	0·61	0·20	0·122	0·10	0·061	0·35	0·214
0	0·64	0·30	0·192	0·20	0·128	0·25	0·160
50	0·66	0·20	0·132	0·20	0·132	0·30	0·198
100	0·68	0·20	0·136	0·30	0·204	0·10	0·068
	Expected utility		0·641		0·643		0·640

It is interesting to note that if the sums involved in this example were much smaller, the decision-maker might well then act in accordance with expected cash value. If we rework this problem (Table 7) with all the possible cash values divided by 10, we obtain Table 9.

The difference between the investments on the utility scale is small, but this is only a reflection of the fact that we chose the zero and unit of the scale as − £1000 and + £1000 with certainty, respectively. The analysis definitely ranks the investments in the order B, A, C, i.e. the same as that of their expected cash values. This is in keeping with known behaviour that, when the sums involved are relatively small, less importance is attached to the risk involved when making a decision. It can be explained in terms of utility theory by noting that over a small range the utility curve can be approximated as a straight line; if the utility curve is a straight line then a decision-maker will act on the basis of expected cash value. This also explains "underwriting". To an individual, investment B is not an acceptable proposition; to a consortium of 10 people, each sharing the risks and pay-offs equally, it becomes worth while.

REFERENCES

[9] R. D. Luce and H. Raiffa (1957) *Games and Decisions.* Wiley, New York.
[12] J. von Neumann and O. Morgenstern (1944) *The Theory of Games and Economic Behaviour.* Princeton University Press, Princeton.
[13] H. Chernoff and L. E. Moses (1959) *Elementary Decision Theory.* Wiley, New York.
[14] P. E. Green (1963) Risk attitudes and chemical investment decisions. *Chem. Eng. Progr.* **59**, 35.

This extract makes two closely related points. Early on, Adelson emphasizes that utility functions do not just encode a decision maker's preferences; they encode his preferences in the context of risk. In his words, they provide 'a portrayal of an individual's attitude to risk'. It is this point that we were making, albeit in different words, when we emphasized that the utility function must be assessed in a manner that ensures that expected utilities provide an appropriate ranking of the actions in his decision problem.

Later in discussing the assessed utility curve shown in his Figure 2, Adelson notes that it 'reflects a conservative attitude towards risk taking'. In general, all concave (bending downwards) utility curves do this (see e.g. French, 1986). A decision maker is **risk averse**, i.e. behaves conservatively, if his utility function is concave. For concave $u(\cdot)$ for any x and x':

$$u[(x + x')/2] > (1/2)\cdot u(x) + (1/2)\cdot u(x')$$

So a risk-averse decision maker prefers to receive $(x + x')/2$ pounds for certain than to take part in a 50–50 gamble in which the two prizes are x and x' pounds, whatever the numerical values of x and x'. A decision maker is **risk prone** if his utility function is convex (bending upwards). A risk-prone decision maker prefers to take part in the above 50–50 gamble than to receive $(x + x')/2$ pounds for certain. As Adelson remarks in his footnote, a decision maker's utility curve need be neither concave nor convex. It may concave in some regions and convex in others, reflecting differing attitudes to risk depending on the magnitude of the sums of money involved.

If his utility function is linear, say

$$u(x) = cx + d$$

a decision maker exhibits a **risk neutral** attitude. He is indifferent between accepting for certain the EMV of a gamble and taking part in the gamble. In particular, because

$$\begin{aligned} u[(x + x')/2] &= c(x + x')/2 + d \\ &= (1/2)\cdot(cx + d) + (1/2)\cdot(cx' + d) \\ &= (1/2)\cdot u(x) + (1/2)\cdot u(x') \end{aligned}$$

he is indifferent between receiving $(x + x')/2$ pounds for certain and a 50–50 gamble with prizes x and x' pounds. Since $(x + x')/2$ pounds is the EMV of the gamble, a risk-neutral decision maker may rank the alternatives in his decision problem according to EMV instead of expected utility. In most cases decision makers' utility functions are not linear. For small ranges of x, however, they may be approximated by linear functions; see Figure 2.4.

Of course, what is meant by a 'small range' depends on context. For an individual considering his personal finances it may be only £10 or £20; but for a large corporation with substantial assets it may be several hundred thousand pounds. Here we have the justification for using EMV as the

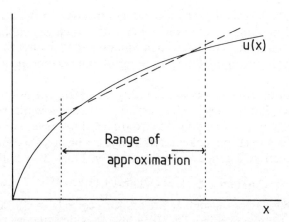

Figure 2.4 A linear approximation to a utility function over a 'small range' of x. In fact, particular values of the parameters c and d need not be found, since ranking actions according to (cEMV + d) gives the same ordering as ranking them according to EMV alone.

criterion in the Rev Counter decision: cf. Mr Pakin's comments at the beginning of Part III of that study. Note also Adelson's comments at the end the extract.

One usually talks of a decision maker's attitude to risk only in cases where the consequences are purely monetary. All that is necessary, however, is that the consequences can be described in some unidimensional way: they might be times of journeys, number of jobs created, etc.

The parallels between the assessment of a subjective probability distribution and that of a utility function should be apparent. In discussing subjective probability assessment we made several comments; and many of those comments are equally appropriate here.

First, the decision maker is not expected to identify precisely a sector of angle z degrees on a probability wheel such that he is indifferent between a gamble between the best and worst consequence and receiving a consequence for certain. All that the assessor asks is that he locates a range of z for which he is indifferent. Later in analysing the decision tree, the significance of the decision maker's lack of discrimination can (and should) be investigated.

Second, the assessment of a utility function encourages a decision maker to think about his preferences and whether they are consistent. Suppose you underwent the assessment described by Adelson. Would you be able to answer the questions about your preferences between gambles without considering how much of the EMV of a gamble to trade off to avoid its risk? Since you are required to answer several questions in the same vein, you must surely explore and come to understand your preferences better. Moreover, and this point is not brought out by Adelson, there is usually some consistency checking built into the assessment. For instance, in Adelson's example, you might be asked which of the following bets you prefer.

Bet C: £300 with probability 0·4;
          −£650 with probability 0·6.
Bet D: a loss of £200 for certain.

Since

$$0·4u(300) + 0·6u(-650) = 0·5$$
$$= u(-200)$$

you should, if consistent, be indifferent. If you do have a strong preference, then the assessor must pause and ask you to reflect upon your earlier judgements and revise them so that they are consistent. As we have noted, it is an assumption of the Bayesian school that you would be willing and able to make the necessary revisions.

The above does little more than sketch the procedure for assessing a utility function. There are many more subtle points that might be made and many variants of the procedures that might be discussed. I shall simply refer you to the literature; see, *inter alia*, Farquhar (1984), Hull *et al.* (1973), Keeney and Raiffa (1976), and von Winterfeldt and Edwards (1986). As in the case of the assessment of subjective probabilities, the assessment of utilities must be done in such a way as to guard against psychological bias (McCord and de Neufville, 1983).

2.4 MULTI-ATTRIBUTE UTILITY THEORY

Despite my remarks in section 2.1 that the consequences $x_{i,j}$ in the representation of a decision problem are descriptions, not numbers, we have only really discussed the assessment of utility and the analysis of the problem in the case that they are purely monetary. Decision trees have had

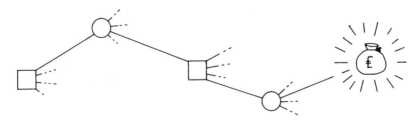

Figure 2.5 A simple decision tree.

the form shown in Figure 2.5: at the end of each branch was the proverbial crock of gold.

True, the method of assessing $u(x_{i,j})$ by comparing $x_{i,j}$ with gambles between x_{best} and x_{worst}, in principle, allows one to determine a decision maker's utility function for any set of consequences; but most decision makers would, in practice, balk at making the necessary comparisons if the consequences were at all complex. Could you, with the aid of a probability wheel or not, identify to within a small range the probability p which would make you indifferent between the following?

> Bet A: a probability p of winning – a three-week holiday in the Bahamas in a five-star hotel with sufficient pocket money to investigate planters' punch fully;
> a probability $(1 - p)$ of winning – a weekend break in January at Mrs Grump's bed and breakfast establishment (parking, TV in every room) in Bognor Regis.
> Bet B: receiving for certain – a fortnight's holiday in Majorca in a three-star hotel on a limited budget.

The consequences of decisions in real life are seldom as simple as even this choice between holidays would suggest. They are complex, involve non-monetary attributes and have to be evaluated in the light of the decision maker's many, usually conflicting objectives. The methods of the previous section will not, of themselves, give us a way of assessing utilities in most real problems. Even when the consequences appear to be entirely monetary, things may not be simple because monetary consequences usually accrue over time rather than arrive as a lump sum. Consider the life-cycle costs of a system shown schematically in Figure 2.6. I have in

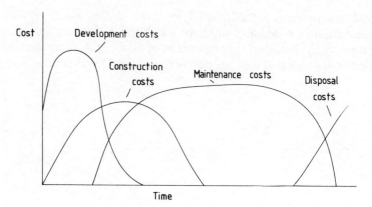

Figure 2.6 The cost profiles of a large system (weapon, transport, etc.).

mind a large weapon system or transport system with a life expectancy of some tens of years.

The usual approach to evaluating such a system is to reduce the entire system cost to a single monetary value by:

1. *Discretizing:* i.e. breaking into yearly (quarterly, monthly, . . .) costs.
2. *Discounting:* i.e. applying a discount factor to bring all costs to 'present values'.
3. *Accumulating:* i.e. adding all the costs at present values together.

If a decision analysis concerned the development of such a system and the only factors that the decision makers considered important were the costs, then the consequences at the ends of the branches of the decision tree might be taken as the net present values of the life-cycle costs under different assumptions about the exogenous economic state. If, further, we make the unlikely assumption that the decision makers' attitude to risk is such that EMV may be used as the decision criterion, then we have a situation that may be modelled in the 'crock of gold' form and analysed by the methods that we have discussed so far.

But is it likely to be sufficient to summarize the entire life-cycle cost profiles by a single number: a net present value? Are the decision makers likely to consider a pound spent on development as equal in value to a pound spent on maintenance? Development engineering may have spin-offs, sustains research potential, fosters company or national pride, etc. Maintenance engineering has fewer by-products. We might ask whether smoothness of profiles is important. It is quite possible for the two maintenance cost profiles shown schematically in Figure 2.7 to have the same net present value. Yet most managers would have a strong preference between them – usually for the smooth, steady profile.

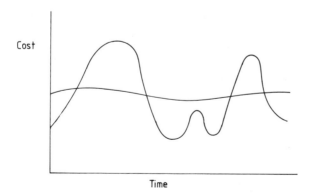

Figure 2.7 Two possible maintenance cost profiles.

Moreover, discounting and uncertainty do not always go together happily. Which of the following bets based on a fair coin do you prefer? (x, y) represents a salary of x pounds this year and y pounds next year.

Bet C: (£30,000, £15,000) if heads;
        (£15,000, £30,000) if tails.
Bet D: (£30,000, £30,000) if heads;
        (£15,000, £15,000) if tails.

In Bet C there is uncertainty about the distribution of your income over the two years, but the certainty of a total income of £45,000. In Bet D the distribution of your income is certain: you will receive equal amounts in both years. It is your total income which is uncertain. I am pretty sure that you would have a preference between the two bets. Yet, whatever discount rate you choose, it is a simple matter to show that the two bets have the same expected net present value.

The import of these remarks is, I believe, that net present values may not reflect decision makers' preferences for the consequences in their problems sufficiently well for decision analyses based upon them to be valid. Furthermore, the problem is not simply one of attitude to risk. Net present values are too condensed a summary of cost profiles. They ignore information about the temporal distribution of income and expenditure which is a significant determinant of decision makers' preferences.

There is in these remarks implicit criticism of the analysis described by Moore and Thomas in section 1.2. It should be remembered, however, that their paper was published in 1973 and referred to work done a year or two before that. Disquiet with net present values in decision analysis is more recent. Moreover, the suggestion is that net present values may be inappropriate: not that they necessarily are for all analyses.

Given that we have doubts about how to evaluate monetary consequences, how can we hope to deal with non-monetary ones? Fortunately, the problem is not as horrific as it seems. We began is section 2.1 with $x_{i,j}$ being a complete verbal description of the consequence of taking action a_i when s_j was the true state of nature. Then imperceptibly we allowed ourselves to summarize this description into a single number, a monetary value. Our difficulty is not that we summarized the description, but that we summarized it too far.

The decision makers may need several summary statistics – or **attributes**, as they are known in decision analysis – to describe complex consequences adequately. Thus we shall assume that the consequences are represented as vectors of attribute levels. The idea is that the attribute levels measure the degree of success or achievement, in the decision makers' opinions, of a consequence against those objectives or factors which they consider to be the prime determinants of their preferences. Of course,

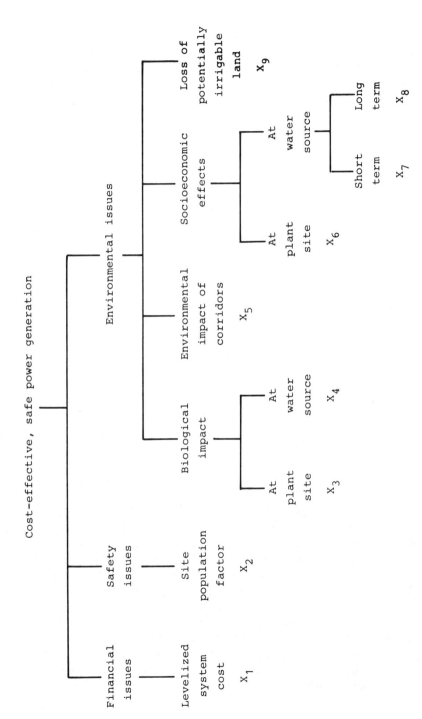

Figure 2.8 Hierarchy of attributes for nuclear power plant site selection.

identifying which attributes to use is not easy: it is also an art more than a science.

Perhaps the easiest way in which to introduce the underlying ideas is to note that attributes are naturally gathered together in hierarchies. Figure 2.8 shows a hierarchy from a case study of a selection of a nuclear power plant site (Kirkwood, 1982). (NB the actual hierarchy was not published in the paper; I infer its form from earlier papers.) As one works down a hierarchy, the level of description of the potential consequences moves from the general to the particular. Moreover, the description is always in terms of factors that are relevant in determining preference. Thus at the top of the hierarchy are the words 'cost-effective, safe power generation', indicating that overall potential sites are to be described and, therefore, ranked in these terms. The next level of the hierarchy indicates that this rather nebulous, overall description is composed of three groups of measures relating to financial, safety and environmental issues; and so on down the hierarchy. There are four environmental issues, one of which is the biological impact due to the power station. In turn, the biological impact has two components: the impact at the site itself, and the impact at the place from which water for cooling is taken.

At the lowest level of the hierarchy, there are nine attributes – dimensions, if you prefer – in terms of which the consequence of choosing each site could be described. They are labelled $X_1, X_2, \ldots, X_9$. Their meaning is explained further in Tables 2.3 to 2.5. Each potential consequence could thus be represented by a vector of nine numbers, describing the levels achieved on each of the nine attributes.

The levels of achievement against some of the attributes, such as levelized system cost and site population factor, were easy to define and assess. Others, however, took substantial effort. It was necessary not only that these attributes be adequate indicators of the underlying effects and issues that concerned the decision makers, but also that information be available to assess each attribute's level at each of the alternative candidate sites. For this reason, several of the attributes were defined subjectively. Panels of experts produced a series of descriptions of archetypal effects. Four example descriptions for the attribute 'biological impact at plant site' are given in Table 2.4. These descriptions were simply labelled $0, 1, 2, \ldots, 12$. Each candidate site was then compared with these descriptions and assigned the level of the one that was judged closest. In the case of the attribute 'environmental impact of corridors', the attribute's definition combined objective and subjective aspects. The objective distances of the various corridors were weighted according to subjective assessments of the types of regions crossed. The weights were assigned by comparing the regions actually crossed with descriptions of archetypal regions; see Table 2.5.

Table 2.3 Attributes for the nuclear power plant site selection study (Kirkwood, 1982, Table 2). The site population factor is a standard measure of population density weighted by the inverse square distance from the plant. Tables 3 and 4 of the study are reproduced here in Tables 2.4 and 2.5

Evaluation measure	Definition
X_1: levelized system cost	Levelized annual cost in year 0 dollars. This includes water-sensitive costs, site-sensitive costs and baseline reference costs.
X_2: site population	Site population factor.
X_3: biological impact at plant site	A 13-point constructed scale, as illustrated in Table 3.
X_4: biological impact at water source	A 3-point constructed scale.
X_5: environmental impact of corridors	A weighted sum of the distances for the electrical transmission and water supply pipeline corridors, using the weighting factors illustrated in Table 4.
X_6: socioeconomic effects at plant site	0 Annual population growth rate due to peak-year construction activities is less than 15%. This indicates that there are population centres near the site with existing infrastructure to serve as a base for the new population influx.
	1 Annual population growth rate due to peak-year construction activities is more than 15%. This indicates that there are no existing population centres of significance; boom-bust development is virtually certain to occur.
X_7: short-term socioeconomic effects at water source	The value of production directly affected by the withdrawal of irrigation due to diversion of water to the power plant, in year 0 dollars.
X_8: long-term socioeconomic effects at water source	For each water source, the scale used is: (aggregate annual personal income per annual quantity of water consumption) × (quantity of water supplied to the power plant from the water source) in levelized year 0 dollars.
X_9: loss of potentially irrigable land	Weighted square miles of cropland that could potentially be retired from use at a candidate water source, using the following weighting factors on area:
	1 cropland currently under irrigation
	1 land highly suitable for irrigation
	0.7 land moderately suitable for irrigation

Table 2.4 Example definitions of attribute levels for *biological impact at plant site*. Each potential site was compared with these definitions by environmental specialists. (Kirkwood, 1982, Table 3)

Level	Definition
0	Removal of 6 square miles having $> 25\%$ of cultivated agricultural use.
4	Removal of 6 square miles of grassland, shrubland or pinyon–juniper habitat that includes $\leqslant 10\%$ riparian or wetland habitat.
8	Removal of 6 square miles of grassland, shrubland or pinyon–juniper habitat within 1 mile of significant actual or potential raptor habitat, and of which $\leqslant 25\%$ is actual or potential habitat for threatened, endangered or otherwise unique species.
12	Removal of 6 square miles of grassland, shrubland or pinyon–juniper habitat within 1 mile of significant actual or potential raptor habitat and including $\leqslant 10\%$ riparian or wetland habitat and $> 25\%$ actual or potential habitat for threatened, endangered or otherwise unique species.

Table 2.5 Examples of weighting factors for water supply pipelines used in the attribute *environmental impact of corridors* (Kirkwood, 1982, Table 4)

Raw mileage	Weighted mileage	Criteria
1	1	Route traversing unpopulated rangeland or utilizing existing industrial corridor. Route not affecting any known endangered species or important limited habitats. Route does not intrude on a 'pristine', historic, culturally significant or archaeologic and palaeontologic resource area.
1	3	Route traversing state or federal forested lands, wildlife management or critical habitat areas. Route traversing ecologically sensitive wetlands.
1	10	Route traversing state or national parks or monuments military bases or military research areas. Route traversing habitats containing unusual or unique communities, endangered species or introduced game species.

This example illustrates, I hope, that much subtlety is needed usually to define the attributes with which to describe the consequences. It also illustrates that consequences need not be represented simply as 'crocks of gold'; non-monetary intangible factors can be introduced into the analysis.

Above, we noted some possible failings of net present value ideas in dealing with life-cycle costing problems. There is a way forward if the different types of future costs are represented by different attributes. Suppose that the decision makers do consider the costs of the system to be the only factors of importance in determining their preferences. If the planning period is q years, they might summarize each consequence as a vector of dimension $4q$. The first q components would be the development costs in each of the q years; the next q, the construction costs; and so on (French, 1983a). Taking each of the $4q$ annual costs as an attribute, we have again a representation of the possible consequences as vectors of attribute levels.

As I remarked above, developing an appropriate hierarchy of attributes for a problem is an art rather than a science. One possible approach is illustrated in my paper 'From decision theory to decision analysis' (French, 1984), reprinted in section 4.3. There it is shown how a hypothetical firm deciding where to site a new warehouse might identify an appropriate set of attributes. In Keeney and Raiffa (1976, Ch. 2) there are many other examples based upon real case studies. Further general advice and examples may be found in von Winterfeldt and Edwards (1986).

Henceforth, we shall assume that all consequences are represented as vectors of attribute levels. The idea is to provide a summary, a description, of each. As yet there is no modelling of preference; at least, in an explicit sense, there is not. The consequences have not been compared in terms of preference. However, in making the judgement that the vector of attributes describes the consequences sufficiently well to determine their preferences between them, the decision makers must have considered some aspects of their preferences.

We began this section by suggesting that most decision makers would balk at making the necessary comparisons to assess a utility function over consequences that were at all complex. Simply summarizing these as vectors of attributes is unlikely to make the comparisons substantially easier. Fortunately, this structuring of the description of the consequences almost invariably enables structure within decision makers' preferences to be identified, and it is possible to capitalize upon this structure to facilitate the assessments of utilities greatly.

Consider the very common, but not universal, feature of preferences embodied in the following statements.

All other things being equal, I prefer more money to less.

All other things being equal, I prefer greater safety to less.

All other things being equal, I prefer a greater market share to a smaller one.

In terms of two attributes (X, Y), these statements correspond to: if attribute Y is held constant, then a greater level of X is always preferred to a smaller one, whatever the constant level of Y. When this holds, attribute X is said to be **preferentially independent** of Y.

Preferential independence between attributes does not always hold. As a trivial example of preferential dependence, consider the conventional preferences for wine with a meal. Red is preferred to white with beef; but white to red with fish. Preferences for the attribute 'wine' depend on the value of the other attribute 'main course'. More seriously, we suggested earlier that smooth timestreams of costs might be preferable to greatly varying ones: see Figure 2.7. This suggests that preference is affected by the closeness of one year's costs to previous years'; preferential dependence is being exhibited.

Here we have defined preferential independence for two attributes, X and Y. Corresponding definitions can be made for the case of q attributes; however, they require the introduction of some detailed notation, which neither is needed elsewhere nor brings with it any pedagogic advantages. Since the discussion of independence conditions in higher dimensions introduces no further concepts or principles of importance, we shall confine our attention here and below to the comforting surroundings of two dimensions. Discussions of independence conditions in higher dimensions are, of course, available elsewhere: Bunn (1984), French (1983b, 1986), Keeney and Raiffa (1976), and von Winterfeldt and Edwards (1986).

The importance of recognizing when attributes are preferentially independent is that in such cases the form of the utility function is greatly limited. If attributes X and Y are mutually preferentially independent, then $u(X, Y)$ must have the form:

$$u(x, y) = r(v(x) + w(y))$$

where

$r(\cdot)$ is a strictly increasing, real valued function which encodes the decision maker's attitude to risk;

and

$v(\cdot)$ and $w(\cdot)$ are functions which together encode the decision maker's preferences concerning trade-offs between the two attributes.

(As a matter of notation we use upper case for the attributes themselves and lower case for particular levels or realizations of them; cf. the standard notation for random variables.)

The functions $v(\cdot)$ and $w(\cdot)$ can be assessed independently of the function $r(\cdot)$, thus separating two difficult tasks for the decision maker. He can first consider how he would trade off an increased level of one attribute for a decreased level of the other, without being confused by any considerations of his attitude to risk. Then he can reflect on his attitude to risk without the complicating factor of needing to think about trade-offs. Structure within the decision maker's preferences implies that his utility function must have a certain structure, and this structure enables the assessment to be separated into a number of easier steps. The decision maker can focus on certain aspects of his preferences without being confused by others.

Preferential independence is the simplest of independence conditions. Other conditions can further structure the utility function and, hence, its assessment. For instance, consider the four bets shown in Figure 2.9. They are also illustrated in an alternative form in Figure 2.10. The prizes (X, Y) consist of two sums of money: £X received immediately, £Y received in a year's time.

Note that in bets E and F the second year's prize is fixed at the common level £300; there is no uncertainty about Y. In choosing between bets E and F, it is the first year's prizes that distinguish the outcomes. Similarly, in choosing between bets G and H, there is again no uncertainty about Y: it is the first year's prizes that distinguish the outcomes. Moreover, the uncertainty inherent in the first year's prize for E is the same as that for G; likewise F and H share the same uncertainty. It is not unreasonable to suggest, given the similarity between the bets, that a decision maker would prefer E to F if and only if he preferred G to H.

If, in general, the decision maker's preferences between bets involving a fixed level y of attribute Y and varying levels of attribute X do not depend

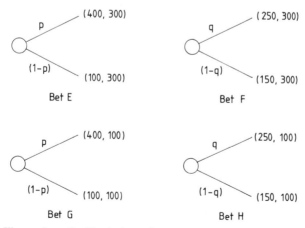

Figure 2.9 Illustration of utility independence: p and q are probabilities.

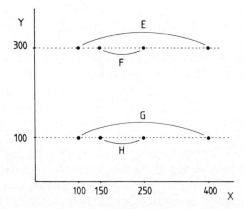

Figure 2.10 Another illustration of the utility independence example.

upon y, then his preferences for X are said to be **utility independent** of Y.

Of course, utility independence need not hold. In this example, had the second year's prizes been more disparate, then it would be a very suspect assumption. Suppose that the fixed second year's prize had been £30,000, not £300, in bets E and F, but that it remained £100 in bets G and H. Suppose also that $p = q = 0.5$. Then it would be quite reasonable for the decision maker to argue that he preferred E to F because, even though it was riskier, with £30,000 guaranteed the following year he could afford to take the risk. Equally, with only £100 guaranteed the following year he might prefer the less risky H to G.

If X and Y are mutually utility independent, then the utility function must have the form:

$$u(x, y) = v(x) + w(y) + kv(x)w(y)$$

where

    $v(\cdot)$ is a utility function representing the decision maker's preferences for bets involving a fixed level of Y and varying levels of X;

    $w(\cdot)$ is a utility function representing the decision maker's preferences for bets involving a fixed level of X and varying levels of Y;

and

    k is a constant.

With a little algebraic manipulation, it may been shown that

$$1 + ku(x, y) = (1 + kv(x))(1 + kw(y))$$

For this reason, $u(x, y)$ is often said to have a **multiplicative form.**

Here again structure within the decision maker's preferences leads to structure in the form of the utility function. Just as we remarked above, this structure helps simplify the assessment of $u(\cdot, \cdot)$. It allows the confounded problems of attitude to risk and trade-offs between the attributes to be separated. In this case $v(\cdot)$ and $w(\cdot)$ can be assessed separately without consideration of trade-offs with the other attribute (because the other attribute is held constant). Then asking the decision maker about his indifferences in the absence of uncertainty provides sufficient information: (i) to ensure that $v(\cdot)$ and $w(\cdot)$ are assessed on the same scale, and (ii) to determine k. Further details of the assessment of $u(\cdot, \cdot)$ may be found in Bunn (1984), and Keeney and Raiffa (1976).

Preferential and utility independence are only two of the independence conditions that might hold in a decision maker's preferences. There are many other more subtle conditions. Moreover, dependence conditions have also been studied. These identify a certain structure in the dependence of the preference for one attribute on the level of other attributes. For discussion of all these conditions we refer to the literature: Farquhar (1981); Farquhar and Fishburn (1981), French (1984), and Keeney and Raiffa (1976).

Different sets of independence or dependence conditions lead to different functional forms of the utility function; but all these forms have one feature in common: the utility function over q attributes is constructed from functions of many fewer than q attributes; often functions of single attributes. This structuring of the utility function greatly simplifies its assessment in the ways indicated above. The decision maker is able to concentrate on certain aspects of his preferences without needing to consider other aspects simultaneously.

To summarize: in analysing a decision problem with complex consequences, the first step is to represent each consequence as a vector of attribute levels. Next, appropriate independence or dependence conditions are identified. Keeney and Raiffa (1976) indicate how this may be done. Essentially, the decision maker is asked about his preferences in simple choices such as between bets E, F, G and H above. When he comes to appreciate exactly what a particular condition is saying, he is asked whether he believes that his preferences should obey it. With appropriate conditions identified for the context of his problem, the form of the utility function is determined and its values assessed.

One area where the theory is less well developed than perhaps it should be is that of preferences over timestreams. As the early examples in this section indicate, it may not be sufficient to rely on discounting methods. Some work has been done on identifying more appropriate forms of utility function; i.e. on identifying appropriate independence or dependence conditions. However, more needs to be done, although one should not imagine that the task will be easy. In French (1986) I survey the recent literature and indicate what the problems are; Meyer's chapter in Keeney and Raiffa (1976) is a very useful reference.

3

Case studies

3.1 INTRODUCTION

The six case studies in the following sections illustrate the application of the theory discussed in the previous pages. Before commenting on and highlighting particular aspects of these studies, I should like to make two general comments.

First, over the past sixty pages or so, I have presented only the barest bones of the theory; sufficient for the case studies, but little more. There are many other aspects of the theory, to which you will find an occasional allusion in the following pages. For discussion of these, you must go to the literature, cited both in the studies themselves and in the suggestions for further reading at the end of these notes.

Second, my style can be dogmatic at times: some say, bigoted. I write from an ivory tower, from which a world of black and white can be clearly seen. Practical men inhabit a world of shades of grey, where pure theory has to be tempered with expediency and compromises made. For instance, as Beattie says in the first study (section 3.2):

> Although there are disadvantages in equating utility and expected net present worth, in the practical situation it is necessary not only to adopt a method that represents the 'true' utility of each situation but also to convince one's accounting colleagues that it does so.

I fully recognize this, despite my comments in section 2.4. Indeed, one of the many reasons for including case studies in these notes is to make precisely that point. So, if elements of the theory are not applied – or apparently contradicted – in particular studies, do not be surprised. Doubtless, there were practical reasons for doing so. I would claim, however, that the general approach embodied by all these case studies is precisely that which motivated and underlay all the theory of the previous pages.

3.2 MARKETING A NEW PRODUCT

This paper by Beattie was published in 1969. It was, I think, the first case study of decision analysis published in *Operational Research Quarterly*. Its publication was timely, because in the same issue of the journal Adelson and Norman had recorded that they could not find any published case studies, only 'examples'.

Several points are noteworthy. First and foremost, the output of the analysis was in large part the better understanding of the problem. For instance, at one point Beattie remarks that

> quite a lot was learned about the effect of possible non-optimal policy decisions and the cost of a range of alternative decisions, in addition to the best course of action being indicated.

Throughout, understanding was fostered by a careful and skilled use of sensitivity analysis. The significance of using particular values of parameters and simplifying assumptions in relation to the indication of the optimal choice was constantly checked.

Second, although the model was simple, it was sufficient for the purposes of the decision. In Phillips's terminology (section 3.6), it was 'requisite'. We shall return to this point in section 4.2.

Third, Beattie compares the results of prior analyses with those of preposterior or Bayesian analyses. The distinction being referred to here is that we discussed in section 2.2. We remarked there that all the probabilities associated with branches in a decision tree could be assessed directly (prior analysis). Alternatively, some could be derived from other prior probabilities and likelihoods through Bayes' theorem (preposterior analysis).

Fourth, a point that probably does not need making these days: Beattie remarks that these analyses were not quick to conduct. They took some 4 weeks. Modern microcomputers have removed this time constraint. Decision trees can be built and analysed in a few hours, no longer than the time necessary for the decision makers to think deeply about their choice. A growing practice today is that of **decision conferencing**, in which all the interested parties in a decision, the problem owners, gather together for one or two days, supported by a team of decision analysts and their microcomputers. The time required to make the decision is now almost entirely that required for discussion. The building of the decision tree, the assessment of probabilities and utilities, and the sensitivity analyses are important catalysts of that discussion. We shall discuss decision conferencing in greater detail in section 4.4.

Marketing a New Product

D. W. BEATTIE
Cadbury Group Ltd.

Some practical experiences in collaborating with a marketing department in deciding whether or not to test-market before launching a new product. The benefits derived from the use of elementary decision-tree theory are described.

INTRODUCTION

The company was considering marketing a new product in a section of the highly competitive food market. Consumer tests on samples of product from a small pilot plant had given favourable results. The decision had then to be taken either to put the product on national sale as quickly as possible or to proceed more cautiously by test marketing in a limited region first, to get a better indication of the likely national sale. The former course involved risking large expenditures on plant and promotional activities; the latter approach allowed the collection of additional information before deciding whether to incur the major part of the expenditure, but increased the time before it would become an established profit-making product.

In trying to relate their information to the marketing decision to be made, our market research colleagues felt that there must be a logical way to make the decision. Accordingly, it was decided to try to formulate the problem in terms of a mathematical model.

BASIC DATA

The pilot plant was capable of making up to 15 tons/week of the product and the main plant would be capable of making up to a maximum of 120 tons/week, although the capacity working conventional shifts was somewhat less than this. The main plant would cost £100,000 installed and could be in production within 1 year of the decision to purchase it being made. Initial forecasts were for sales of 70 tons/week (nationally) and the estimators attached a standard error of 20 tons/week to this estimate.

Marketing considerations meant that 6 months had to be allowed between a decision to launch the product in an area and the introduction date. It was estimated that a test market of at least 6 months covering part of the country had a three to one chance of giving a reliable estimate of the national sales level. Likewise the level of sale in the first year of national sale had a 90 per cent chance of giving a reliable estimate of the ultimate sales level. In the event of the test market showing the product to be a failure, it would be possible to restyle the product and be ready for a national re-launch 2 years later. A sales estimate was also given for the restyled product.

Net contributions (after tax) at various levels of sales for each year forward were also available.

BUILDING A MODEL

The first requirement of the model was that it should be capable of solution within the time available. The second was that it should be simple enough for both marketing and operational research personnel to understand and operate it. Therefore a decision tree was drawn to illustrate the various decisions and possible outcomes.

There were two main alternatives: either to launch nationally as quickly as possible; or to test market in one area only, then decide whether to go national as soon as possible thereafter, to drop the idea, or to restyle the product entirely.

In constructing the tree, the first of several arbitrary decisions had to be made. The sales estimate was given as a mean and a standard error (and was assumed to be distributed normally). For decision purposes, however, a continuous range of outcomes had to be arbitrarily represented as a number of discrete outcomes. In practice the distribution was partitioned into three segments and the probability and mean value attaching to each segment were calculated.

The next problem was to assess the utility of each course of action. In this case the criterion adopted was the net present worth. Although there are disadvantages in equating utility and expected net present worth, in the practical situation it is necessary not only to adopt a method that represents the "true" utility of each situation but also to convince one's accounting colleagues that it does so.

Having represented all the outcomes in the decision tree, the probabilities attaching to each were estimated. The first approach (favoured by our marketing colleagues) was to assess *all* the probabilities subjectively and then evaluate each branch of the decision tree by prior analysis. (This method is described by Magee[1] and is illustrated in Figure 1 which shows one section of the decision tree.)

Unfortunately as Adelson and Salkin[3] have pointed out, the subjective assessment of correlated probabilities is notoriously difficult and correlated

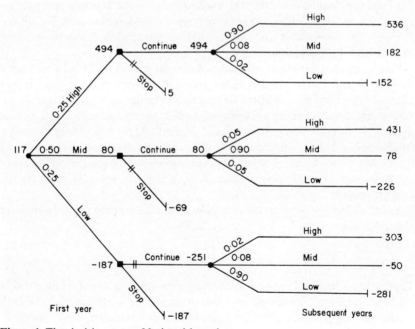

Figure 1 The decision tree – National launch.

probabilities abound in this decision tree. Moreover, useful information on the reliability of test markets as indicators of final sales had not been used explicitly.

As an alternative then, the probabilities were calculated by preposterior analysis. The resulting decision tree will be referred to as a "Bayesian" decision tree. The use of the method in making this type of marketing and investment decision has been described by many authors (e.g. Magee,[2] Langhoff,[4] Green and Frank[5] and Harris[6]). Green and Frank, in particular, give a clear description of the method of calculation, so the calculations will not be presented here.

SOLUTION AND COMMENTS ON THE RESULTS

Using the prior model, the expected net present worth of the two alternatives were:

    A. National launch: £117,000 less £100,000 (capital outlay),
                             i.e. £17,000.
    B. Test market: £133,000 less £91,000,[†]
                             i.e. £42,000.
    Advantage of B over A is £25,000.

Thus the decision indicated was to test market and only be committed to purchase the plant if a sale in the "high" region were indicated.

Although the Bayesian decision tree gave slightly lower values for both alternatives, the advantage of one over the other was virtually the same.

        A. National launch: £12,000.
        B. Test market: £38,000.
        Advantage of B over A is £26,000.

The main difference between the two results was that the prior analysis discounted some of the advantages of test marketing relative to the other model. Under the national launch option, the prior model appeared to overestimate the likelihood of complete success whilst under the test market option it underestimated this chance.

At this point it was interesting to note the effect of the test market. On both models a national launch would only follow the test market if sales in the test market were high. If, however, for "policy reasons" (e.g. maintaining market share, diversification, completion of a range, etc.) one were not prepared to kill the product if sales in the mid region were indicated, this would reduce the value of the test market option. The reduction in net present worth would be £12,000 on the prior model and £15,000 on the Bayesian, but with test marketing still indicated. In this instance, therefore, the first decision was reasonably robust to a subsequent decision being taken in what we have defined as a non-optimal manner. In

---

[†]The capital outlay has been discounted to allow for its being incurred one year later in the test market case.

addition to the analyses already described, a further analysis was made using a series of modified sales estimates to discover the expected level of sale such that each alternative was equally valuable. It was then agreed that to expect a level of sale above this was to be unreasonably optimistic and the decision to test market was confirmed.

From this discussion it will be seen that quite a lot was learned about the effect of possible non-optimal policy decisions and the cost of a range of alternative decisions, in addition to the best course of action being indicated.

PRACTICAL POINTS

Perhaps the most useful thing to emerge from the exercise was the effective interplay of marketing, market research and operational research ideas. This certainly had a beneficial effect on the quantitative minds and, hopefully, on the non-quantitative also. What had been regarded as one large single decision was broken down into a series of smaller decisions and the relationships between these decisions recognized. Thus the subjective information was refined by the asking of questions that would otherwise have remained unasked. Moreover, the building of the models tended to lead to the use of all the available information rather than the very natural tendency for one single factor to disproportionately influence the final decision.

Conventional wisdom suggests that test marketing should always be done. Although the case study confirmed this *in this instance*, it can readily be seen that changes in circumstances (e.g. a higher sales estimate) could easily indicate that the loss of revenue in the gap between test market and national extension – to say nothing of competitive activity – could indicate launching a product nationally as quickly as resources permit.

Lest one is carried away by the technique, one must, however, recognize that to examine a decision in the way described takes time (about 4 weeks in this case) and involves a considerable amount of work. It can only be justified in the case of certain decisions. First, these should be significant decisions. Secondly, they should be ones where the range of alternatives can be considered meaningfully (without having to model *all* the company's activities). Thirdly, they should be capable of suffering sufficient delay to do the necessary work, and, finally, they should be ones where some useful information exists. The decision discussed in this paper for the most part fulfilled these conditions, but it would be idle to suppose that all such decisions will be, or even should be, subjected to this detailed approach. For this reason the benefits of this approach and the interplay of ideas may spread subsequently beyond these isolated cases.

IMPLEMENTATION

Evaluating a decision tree is not the same as making a decision. Fortunately the whole approach to this problem has been worked out from the beginning with market research and the brand manager. Being to this extent committed to the model, committal to the decision followed, despite the earlier feeling that an immediate national launch was the best plan. Thus the decision tree approach proved helpful not merely as a "black-box" decision-maker, but also as a rational way of relating marketing judgements to the sequential decisions which had to be made.

The product has now been test-marketed, launched nationally and will, we trust, continue as an established brand for many years. It would be unwise to assess the goodness of the original subjective probabilities as one is, after the event, in an essentially "one–nought" situation. However, the current level of sales is within one standard error of the original estimate.

CHOICE OF THE MODEL

The models used in this study were unashamedly simple. Other workers in this field have developed more complex models to try to represent the decisions more realistically.

Hertz[7] suggests a simulation approach based on estimates of the *components* of the sales estimate (price, market share, promotional activity, etc.). However, this involves replacing the existing (and trusted) method of making sales estimates by a "black box". His other main idea, the selection of projects on the basis of their risk profiles rather than the expected value only (or on the basis of the mean and standard deviation of the net present worth as suggested by Van Horne[8]) is, of course, possible without recourse to simulation. The even more complex DEMON model of Charnes *et al.*[9] was rejected for similar reasons.

Decision-tree models, on the other hand, are simple and lend themselves particularly well to construction (and modification) in a group approach. In this way the role of the operational research worker is as much to act as a catalyst among the marketing and market research men as to act as a producer of quantitative bases for decisions. Unlike the more complex models, the relationship between these models and the problem is obvious and the element of blind faith correspondingly reduced.

However, simple models do involve making some arbitrary assumptions in order to represent complex problems. Two will be mentioned here; the reader will note several others. First, in partitioning the sales estimate into three discrete events an arbitrary decision was made. Fortunately, analysis of the model with different three-way splits and one four-way split yielded the same decision (but the advantage of the test market ranged from £23,000 to £37,000 with the prior model, and from £26,000 to £34,000 with the Bayesian model). Secondly, the choice of planning horizon was arbitrary. The net present worth was calculated on an assumed product life of several years: the stop–go decision process *in the model* terminated in the second year of national sale. In reality, opportunities to withdraw the product are always present as are other possibilities (e.g. withdraw advertising support, change the price, etc.).

CONCLUSION

Despite the objections to the use of decision trees on the grounds that they do not completely represent the situation being modelled, in this case study they proved useful in making a decision which would otherwise have been made entirely by hunch. Until such times that means of obtaining better data and more sophisticated but practicable models are available, the contribution that operational research can make to the general problem of decision-making under uncertainty will be subject to these sorts of limitations. Nevertheless, if the problems exist, the role of the operational research workers should be to help those who *must* obtain a solution to them.

ACKNOWLEDGEMENTS

Thanks are due to the Directors of Cadbury Group Limited for permission to publish this paper; also to colleagues in the Market Research Department without whose help and initiative the case study might never have happened.

REFERENCES

[1] J. F. Magee (1964) Decision trees for decision making. *Hvd Bus. Rev.* July/Aug., 126.
[2] J. F. Magee (1964) How to use decision trees in capital investment. *Hvd Bus. Rev.* Sept./Oct., 70.
[3] P. M. Adelson and G. R. Salkin (1967) A Survey of Risk Theory. Paper presented to the Operational Research Society Conference, Exeter, September, 1967.
[4] P. Langhoff (Ed.) (1965) *Models, Measurements and Marketing*, pp. 182–197. Prentice-Hall, Englewood Cliffs.
[5] P. E. Green and R. E. Frank (1966) Bayesian statistics and market research. *Appl. Stat.* **15**, 173.
[6] L. Harris (1967) New product marketing: A case study in decision making under uncertainty. *Appl. Stat.* **16**, 39.
[7] D. B. Hertz (1968) Investment policies that pay off. *Hvd Bus. Rev.* Jan./Feb., 96.
[8] J. C. Van Horne (1969) The analysis of uncertainty resolution in capital budgeting for new products. *Mgmt Sci.* **15**, 376.
[9] A. Charnes, W. W. Cooper, J. K. Devoe and D. B. Learner (1968) DEMON Mk II: An extremal equation approach to new product marketing. *Mgmt Sci.* **14**, 513.

3.3   BALANCING FAILURES IN PACEMAKERS

Ronen, Pliskin and Feldman's paper dates from 1984. It is not, in fact, a case study in the true sense of 'case study': it does not refer to a particular decision problem that occurred and was solved by decision analysis. Rather, it is an 'example' in the sense used by Adelson and Norman (1969). Its purpose is to show how decision analysis can help in the design of a cardiac pacemaker. None the less, it is not a hypothetical example. It is based upon real data: utilities were assessed, and conclusions were drawn.

Many of the points that I made about Beattie's paper are equally apt here. In particular, again the import of the analysis is, in large part, understanding. For instance, note the last paragraph of their paper. They remark that, although the analysis can identify a better design of pacemaker, more crucially, through bringing an understanding of each design's weak spots, it can contribute to the design process itself.

The importance of sensitivity analysis in bringing the user qualitative understanding is again clear.

The assessment of utilities is illustrated and the important point made that the questions asked of the decision maker must be framed in a language that is meaningful to him.

Although our discussion in section 2.2 emphasized the assessment of subjective probabilities by the use of bets based on a probability wheel or some such device, there is no requirement to use such approaches. If past statistical reliability analyses are available, as they are in this study, then the probabilities that they provide can and should be used in analysing the decision tree.

Balancing the Failure Modes in the Electronic Circuit of a Cardiac Pacemaker: A Decision Analysis

BOAZ RONEN*, JOSEPH S. PLISKIN[††] and SHLOMO FELDMANS§
*ELTA Electronics Industries Ltd (a subsidiary of Israel Aircraft Industries Ltd).
†Department of Industrial Engineering and Management, Ben-Gurion University of the Negev and §Heart Institute Chaim Sheba Medical Center, Tel-Hashomer and Sackler School of Medicine, Tel-Aviv University

Cardiac pacemaker malfunctions are of continuous concern to the medical profession as well as to the electronics industry. Certain failures in cardiac pacemaker performance are critical and can result in patient deaths. Cardiac pacemakers are vulnerable to certain malfunctions in the electrode, batteries and the electronic circuit. This paper focuses on failures and reliability of the electronic circuit and how they affect its choice and design. The paper discusses the issues of balancing the various failure modes by considering both failure rates and failure outcomes.

As the choice of an appropriate pacemaker is a decision problem under conditions of uncertainty, we employ decision analysis as the analytical and conceptual framework. Use of utility theory enables a systematic quantitative evaluation of such seeming intangibles as the various failure outcomes. Probabilities are assessed using specific engineering and reliability literature. The method is demonstrated on a choice problem between two specific electronic circuit designs.

‡ Parts of this work were performed while the author was with the School of Management, Boston University, and the Center for the Analysis of Health Practices, Harvard School of Public Health.

The methodology can be useful in designing the electronic circuit to meet certain reliability specifications, deciding whether or not to introduce redundancy, decisions affecting components and technology, and establishing minimal reliability standards, with regard not only to cardiac pacemakers but to other electronics as well.

INTRODUCTION

An electronic cardiac pacemaker is a device that artificially controls the pulse rate of the heart. When the intrinsic pacemaker of the heart cannot provide adequate spontaneous rhythm, there may be a need to implant an electronic pacemaker. It provides electronic stimulation through a special electrode (transvenous or myocardial) that connects the pacemaker to the heart.

Most earlier versions of pacemakers generated a constant electronic stimulation. The newer models provide stimulation 'on demand'. If the intrinsic pacemaker of the heart can function to some degree on its own, the tendency is to let it do so (while the implanted pacemaker 'rests') and to call for the electronic pacemaker stimulation when the intrinsic pace is inadequate. Today's pacemakers have reached a level of sophistication where microcircuit chip construction allows for external reprogramming and telemetry.

Cardiac pacemakers are vulnerable to malfunctions in the electrode, batteries and electronic circuit, and pacemaker failures can also occur owing to ineffective sensing (in demand pacemakers) and infection. This paper will focus on the design phase of the pacemaker and in choices among alternative electronic designs. We shall therefore concentrate on failures and reliability of the electronic circuit. The more general problem of choosing among several pacemakers where all of the above failures are considered is discussed elsewhere.[1]

Cardiac pacemaker malfunctions are of continuous concern to the medical profession as well as to the electronics industries. Certain failures in pacemaker performance are critical and can result in patient deaths. This has resulted in strict governmental standards (in the U.S.A.) to preserve appropriate reliability on the part of pacemaker manufacturers. The highly sophisticated nature of today's pacemakers provides a strong incentive to constantly evaluate their performance and failures. Despite the strict standards issued by the Federal Food and Drug Administration, there still exist problems with several manufacturers.[2,3]

Of 123,000 pacemakers implanted from 1973 to 1975, 22,300 units were labelled as suspicious for malfunctions. All pacemakers must be periodically checked to ensure appropriate pulse-generating frequency. Thousands of patients had to undergo additional surgery to replace defective pacemakers, thus incurring additional costs, risks and inconvenience.

One of the main problems in pacemaker reliability research is failure definition. Every manufacturer, research team and medical institution seems to define failures differently, reflecting their own point of view and interests. This paper tries to define all failure modes clearly and in a unified manner.

Some failure modes are more frequent than others; some have more serious implications for the patient. This paper discusses the issues of balancing the various failure modes by considering both failure rates and failure outcomes. As the choice of an appropriate pacemaker is a decision problem under conditions of uncertainty, we employ decision analysis as the analytical and conceptual framework.[4] Use of utility theory enables a systematic quantitative evaluation of such seeming intangibles as the various failure outcomes.

The methodology presented in this paper can be useful in designing the electronic circuit, deciding to introduce redundancy, helping physicians choose among several pacemakers, and establishing a minimal reliability standard for cardiac pacemakers. The methodology presented allows for a different approach to reliability evaluation in general, especially to systems where failure outcomes are not easily quantifiable.[5]

FAILURES IN THE ELECTRONIC CIRCUIT

A failure in the electronic circuit usually requires replacement of the defective pacemaker with a new one. Several failure modes are possible:

(1) *Runaway pulse* – the pulse frequency of the circuit is faster than originally planned. If the resulting heart rate is far beyond physiologic heart rate, the implications for the patient are very critical and can lead to death.
(2) *Slow pulse* – the pulse frequency is slower than originally designed. The source of this failure can either be the electronic circuit or a reduction in battery voltage. We shall concentrate on the former, where the malfunction is usually caused by a defective component. In most cases, the implications for the patient are not severe. If the failure is detected in time, the pacemaker can simply be replaced without too high a risk for the patient.
(3) *No output* – the pacemaker ceases to provide electronic stimulation.
(4) *Intermittent pulse* – the pacemaker is not properly functioning on a continuous basis. This failure is rare and will not be categorized separately.

Failures due to the electrode, infection or ineffective sensing (in a demand pacemaker) will not be considered in this paper as we are addressing only the electronic circuit. As mentioned earlier, these failures were considered in a broader analysis.[1]

The specific categories we shall consider in the analysis are presented in Table 1. We shall actually not relate to the 'intermittent pulse' failure because of its rarity. In the related study,[1] of the 120 failing pacemakers, not a single 'intermittent' failure has been detected.

THE DECISION PROBLEM

The focus of this paper is the choice problem among alternative designs of the electronic circuit of a cardiac pacemaker. The following reasonable assumptions will underly the analysis:

Table 1. Failure modes and possible outcomes

Failure No.	Failure mode	Outcome
1	No pulse	Pacemaker not functioning at all
2	Exit in high (HI)	Constant flow of current to heart, possible fibrillation
3	Slow pulse	Pacemaker only partially functioning
4	Runway pulse	Pacemaker working too fast
5	Intermittent pulse	Pace changes, interruptions
6	OK	Pacemaker is functioning well

(1) At the design stage there are at least two alternatives (otherwise there is no choice problem).
(2) There are no cost differences among the various alternatives (hence the cost attribute is not relevant).
(3) The reliability of the electronic circuit is determined only by the components (although reliability can be influenced by manufacturing and packaging techniques[5]).

The choice problem enables the decision maker to decide on a specific design but not to control for the actual type of failure whose probability distribution represents the underlying uncertainty in the decision process. Before proceeding with the analysis, we must clarify the issue of whose decision problem we are considering. Is it the designer's, the physician's, the hospital's, or perhaps the patient's? At a first glance we may be tempted to think of the designer as the decision maker regarding the choice of a preferred design. In actuality, it is the cardiologist who chooses one pacemaker over another, and it is his or her subjective preferences that are considered in the choice problem. In theory, it is the patients who must make the choice since the various failure outcomes affect them directly. Because patients may not comprehend the implications of the various end consequences resulting from different failures, we may want to think of the patient-physician team as the actual decision maker. In many cases, though, patients delegate complete authority to the physician to act on their behalf, reflecting their (the patients') best interests. On this premise we decided to use the preferences of the cardiologist involved in this study (S.F.), knowing that other cardiologists may express a different preference pattern, resulting in different choices of pacemakers. An underlying premise for any physician's preferences is that these preferences reflect what the physician believes is best for the patient.

For our analysis we have decided to combine failure modes 1 and 3 into a combined category called 'no pulse' because failure 1 is the limiting case for failure 3 as the frequency approaches zero. The two modes are combined because the available data relate to extreme cases (i.e. 'on', 'off' situations).[6] Also, as far as

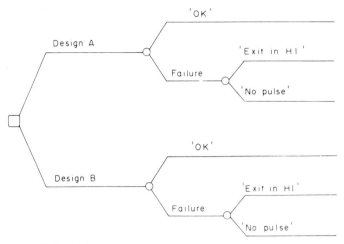

Figure 1 Simplified decision tree.

medical outcomes are concerned, these two failure modes can be similarly detected and have similar implications for the patient. Actually, the patient is rarely in a situation where there is an absolute constant dependence on the pacemaker. The heart functions to some degree, and the pacemaker is there for 'back-up' purposes. Hence, there is little differentiation between these two failure modes. Similarly, failure 2 is the limiting case for failure 4, and they too will be combined into a single category, 'exit in HI'. This failure type implies a constant output of electronic current to the heart and can result in fibrillation.

The decision tree of Figure 1 depicts the simplified decision problem after some of the failure modes have been combined as previously discussed. Note that there are two levels of uncertainty: first for whether a failure occurred at all or not, and then, given that a failure had occurred, there is a chance node for the actual failure mode. The two chance nodes for each of the alternatives can be combined into one, but the representation of Figure 1 is more convenient in terms of the probabilities that need to be assessed.

UTILITY ASSESSMENT

In the representation of Figure 1, there are only three possible outcomes relating to an 'OK' state, 'exit in HI' and 'no pulse'. The 'best' outcome as far as the patient is concerned is obviously 'OK', and the worst is 'exit in HI'. Once the outcomes have been ranked, we can proceed as follows. Two utility values can be assigned arbitrarily. That sets the origin and unit of measurement. It is convenient to set the utility value of the best outcome as 1 and worst as 0. Hence $U(\text{OK}) = 1$, $U(\text{exit in HI}) = 0$. The classical way of assessing the utility of the intermediate outcome, 'no pulse', is to consider the following choice problem: you can use a pacemaker which you know for certain is going to have a 'no pulse' failure, or you could use a pacemaker which has a probability p of yielding an 'exit in HI' failure and a complementary probability $1 - p$ of being 'OK'. Naturally, if $p = 1$ the choice would be for the pacemaker with the 'no pulse' failure. If $p = 0$ the other pacemaker would be preferred. There is some intermediate value p_0 for which the decision maker has a difficult time deciding (i.e. he is indifferent). If so, the utility of the two choices will be the same. The utility of a gamble or chance node is simply the expected value of the utilities of the individual outcomes.[4] The indifference implies

$$U(\text{no pulse}) = p_0 U(\text{exit in HI}) + (1 - p_0)U(\text{OK}),$$

but $U(\text{OK}) = 1$, $U(\text{exit in HI}) = 0$, hence $U(\text{no pulse}) = 1 - p_0$.

As posed above, the choice problem is somewhat unrealistic, and cardiologists find it difficult to think of such hypothetical situations. We revised the approach and used the following choice problem. You are presented with two batches of 1000 pacemakers each. One batch contains 999 defect-free pacemakers and one pacemaker having a 'no pulse' failure. The other batch contains one pacemaker with an 'exit in HI' failure and the rest 'OK'. Which batch would you rather work with? This choice problem is the realistic one faced by the cardiologists and/or the designers. When preferring one design over another, they are in fact choosing one probability distribution over another. Every reasonably sized ordered batch will contain some defective pacemakers, which will eventually be implanted in patients.

Since an 'exit in HI' failure is more severe, most decision makers would choose to work with the first batch (A). But what if batch A contained two defective pacemakers with 'no pulse', or 10, or 20? The choice is no longer obvious. Different decision makers (cardiologists, designers) will exhibit different preference patterns. In general, if the decision maker is indifferent between having a proportion p_A of 'no pulse' defects and having a proportion p_B of 'exit in HI' defects, we can calculate the utility of 'no pulse'. Indifference between the two gambles implies equal expected utility values for the two choices, hence:

$$p_A U(\text{no pulse}) + (1 - p_A)U(\text{OK}) = p_B U(\text{exit in HI}) + (1 - p_B)U(\text{OK}).$$

With $U(\text{OK}) = 1$ and $U(\text{exit in HI}) = 0$, this reduces to

$$p_A U(\text{no pulse}) + (1 - p_A) = 1 - p_B,$$

or

$$U(\text{no pulse}) = \frac{p_A - p_B}{p_A}.$$

The cardiologist involved in this study was indifferent between working with batch A containing 50 'no pulse' pacemakers (and 950 'OK') and batch B containing 1 'exit in HI' pacemaker. We then have

$$p_A = 0{\cdot}050 \quad \text{and} \quad p_B = 0{\cdot}001,$$

yielding

$$U(\text{no pulse}) = \frac{0{\cdot}050 - 0{\cdot}001}{0{\cdot}050} = 0{\cdot}98.$$

As this exhibited preference is highly subjective, other decision makers are not expected to have the same utility structure.

If we wish to consider additional intermediate outcomes, we can assess their utilities in a similar manner. The implications of the obtained utility value will be further discussed in the sensitivity analysis section. Again, it should be noted that this 'batch' approach of eliciting utilities was the only one we found meaningful for the cardiologist.

PROBABILITY ASSESSMENTS

Once we have utility values for all outcomes, we must weight them according to the likelihood of realizing each of the outcomes.

The failure rates can be calculated using the *Military Standardization Handbook 217C*,[6] and the conditional probabilities needed for the second level of the chance nodes of Figure 1 can be obtained from the *Reliability Prediction Notebook*.[7] The calculations involve the following assumptions:

(1) The probability of two or more independent failures in different components occurring simultaneously is negligible.
(2) Components' designations that do not appear on the electronic circuit schemes are evaluated by engineering judgments.
(3) Every component can fail in one of two ways: open circuit or short circuit.
(4) The failure occurrence (open or short circuit) of every component is Poisson with rate λP, where λ is obtained from *MILHDBK 217 C*,[6] Table 2.1.7-3, and P from the *Reliability Prediction Notebook*.[7]
(5) The 'no pulse' and 'exit in HI' failure rates are calculated as follows:

    (a) The effect of every component on circuit output will be checked. It will be determined whether the output is 'no pulse', 'exit in HI' or 'OK'.
    (b) The failure rate, λP, will be calculated for every state.
    (c) The effect of a short circuit in a component on circuit output will be determined.
    (d) The effect of a short or open circuit on all circuit components will be determined.
    (e) The 'no pulse' rate will be calculated as the sum of the λPs of the components and states yielding a 'no pulse' outcome, similarly for the 'exit in HI' rate.

The preferred circuit design can now be determined by calculating the expected utility for each alternative and choosing the one yielding the highest expected utility. An actual example will best demonstrate the procedure.

AN EXAMPLE

Consider the two alternative electronic designs of cardiac pacemakers taken from the literature,[8] as they are presented in Figures 2A and 2B. These two designs were readily available, which is not the case in general. Also, they represent a rather old generation of pacemakers. Current pacemaker designs are generated by computerized methods using large-scale or very large-scale integration circuits. This enables immediate determination of failure modes as well as probabilities of failure. The designs used in our example facilitate a good demonstration of the decision analytic approach. Today's technology makes it easier to use this approach, and hence enhances its value.

At a first glance, pacemaker A (Figure 2A) seems to have more components that render it less reliable, hence a higher failure rate. On the other hand, it contains a capacitor at the exit, thus reducing the probability for the critical outcome 'exit in HI'. Pacemaker B (Figure 2B) is simpler in structure, and hence has an overall lower failure rate. But it has a higher conditional probability of the 'exit in HI' failure.

The calculations yielded:

for Pacemaker A:

$$\lambda P_{\text{no pulse}} = 2{\cdot}6 \times 10^{-8} \text{ failures/hour}$$

$$\lambda P_{\text{exit in HI}} = 2{\cdot}4 \times 10^{-9} \text{ failures/hour}$$

$$\lambda P_{\text{Total}} = \lambda P_{\text{no pulse}} + \lambda P_{\text{exit in HI}} = 2{\cdot}84 \times 10^{-8} \text{ failures/hour;}$$

for Pacemaker B:

$$\lambda P_{\text{no pulse}} = 1{\cdot}9 \times 10^{-8} \text{ failures/hour}$$

$$\lambda P_{\text{exit in HI}} = 5{\cdot}3 \times 10^{-9} \text{ failures/hour}$$

$$\lambda P_{\text{Total}} = 2{\cdot}43 \times 10^{-8} \text{ failures/hour.}$$

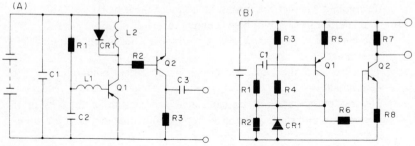

Figure 2 Electronic schemes for two pacemakers. (A) Pacemaker A; (B) Pacemaker B.

Table 2. Probabilities for the two pacemakers

Pacemaker	P(OK)	P(no pulse)	P(exit in HI)
A	0·99950	$4·552 \times 10^{-4}$	$4·20 \times 10^{-5}$
B	0·99957	$3·327 \times 10^{-4}$	$9·28 \times 10^{-5}$

Table 3. Conditional probabilities of failure

Pacemaker	P(no pulse|failure)	P(exit in HI|failure)
A	0·909	0·091
B	0·782	0·218

Mission time was determined as two years, which are 17,520 hours. The failure probabilities needed for the decision tree of Figure 1 are presented in Tables 2 and 3. The expected utility (using the utility values of the previous section) for design A is 0·99994, and for design B 0·99989; hence design A is the preferred one. Again, this preference reflects a single decision maker, and no generalizations should be made. The two expected utility values seem very close, but the actual magnitude of a utility value is meaningless and has value only in comparison to other utility (or

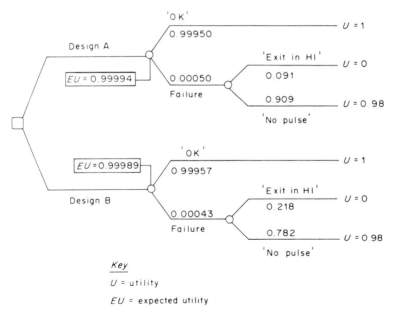

Figure 3 Decision tree and calculations for example.

expected utility) values. The decision tree containing the probabilities and utility values is presented in Figure 3.

SENSITIVITY ANALYSES

As with any decision analysis, it is important to examine the stability of the solution by analyzing its sensitivity to changes in various parameters. From Figure 3 we see that the parameters of concern are the total failure rates for the two designs, the conditional failure mode probabilities, and the intermediate utility value for the 'no pulse' failure. The latter quantity is independent of the two example designs and is a universal figure for the *specific* cardiologist.

(1) *Sensitivity analysis of the cardiologist's preferences*

Specifically, the one quantity assessed by the cardiologist was the utility value of the 'no pulse' failure, U(no pulse). Let us calculate the utility value U_0 for which a shift in the decision from design A to design B would occur, leaving other elements of the decision problem unchanged.

$$EU(A) = U_0(0.909)(0.0005) + (0.9995)(1)$$
$$EU(B) = U_0(0.782)(0.00043) + (0.99957)(1)$$

Equating the two terms, we obtain $U_0 = 0.592$, compared to U(no pulse) $= 0.98$ which we originally used. This represents a substantial difference. Let us observe the implication of this number on the answer to the choice problem described earlier. If we leave $p_B = 0.001$, we obtain via the relationship U(no pulse) $= (p_A - p_B)/p_A$ a value $p_A = 0.00245$. This implies a defective rate of 2.45 per thousand as opposed to the 50 per thousand originally obtained.

There was no way, in the cardiologist's mind, that his indifference value could be that low. From informed communication with other cardiologists we observed that other subjective preferences will also result in a higher proportion than 0.245% of pacemakers having the less severe defect that would offset the one in a thousand pacemakers with the more severe failure.

(2) *Sensitivity analysis of the conditional failure probabilities*

The conditional failure probabilities of Figure 3 were from various 'manual' engineering assessments, and not 'mechanically' by the M.T.B.F. (Mean Time Between Failures) model of *MILHDBK217C*. Therefore, they are the first probabilities to be questioned. In order to achieve a shift from design A to design B, the 'no pulse' probability of design B should increase to 0.878 (from 0.782), assuming all else remains constant, or the respective probability for design A would have to be reduced from 0.909 to 0.796. These shifts represent changes of 12–14% in magnitude, indicating a rather stable solution.

(3) *Sensitivity analysis of the M.T.B.F. model calculation*

We examine the implications of changes in the total failure rate on the optimal decision. First, let us assume that the failure rate on both designs is consistently higher (or lower) than originally used in Figure 3. This is a common engineering phenomenon that could arise, for example, from higher humidity and more fluids in the body environment surrounding the pacemaker. If the failure rate is C, where C is some constant (the same for both designs because we are assuming a consistent error in the M.T.B.F. model), we can determine the value of C that would cause a shift from design A to B. Calculations reveal that such a shift will not occur, regardless of the value of C.

If we examine for possible M.T.B.F. errors in one design only, we obtain that the total failure rate of design A must double to 0·001 for design B to be preferred. This magnitude of error is highly unlikely. If we vary the total failure rate for design B, it must be reduced to 0·0002 from 0·00043. This again is unlikely to be the case.

DISCUSSION

The paper considers only failures in the electronic circuit of cardiac pacemakers. Pacemakers can also fail because of problems in the electrode, battery exhaustion, ineffective sensing and infection. Therefore, the model of this paper is not a decision aid for the cardiologist in choosing one pacemaker over another, because that choice problem must consider all failures. Our model can be very valuable in the design and planning stage of the electronic circuit where cardiologists should interact with manufacturers to provide a desired level of electronic reliability. Given alternative designs, the approach of this paper can serve as a valuable decision aid for choosing a preferred design, deciding whether to introduce redundancy, choosing components, deciding on addition or deletion of components, and choice of technology.

The standard procedure of failure mode error analysis only checks for implication on the total M.T.B.F. of the circuit. Our approach puts the proper weights not only on the probability of failures but also on their criticality.

The model developed in this paper can also provide for setting minimal reliability standards for the electronic circuit. Based on cardiologists' preferences, manufacturers can plan for a certain level of reliability as reflected by total failure rate as well as by the distribution of the various failure modes. For example, the utility analysis can indicate the maximum proportion of 'exit in HI' failure relative to 'slow pulse' failures that could be acceptable.

Such an analysis can greatly enhance the process of designing the electronic circuit of the pacemaker. If we consider, for example, the formal analysis applied to the choice between designs A and B of Figure 2, the analysis could perhaps identify the 'weak' spots in each design and result in a third design, superior to both A and

B. Such a use of the decision analytic framework is perhaps far more crucial than simply choosing one design or the other.

REFERENCES

[1] B. Rone, J. S. Pliskin, S. Feldman and H. H. Neufeld (1979) Optimal choice of implanted cardiac pacemakers. *Proceedings of the VIth World Symposium on Cardiac Pacing.* Chap. 36–1.

[2] Faulty parts afflict pacemakers. *Electronics* (April 1975).

[3] Justice gets tough with pacemaker firm as companies seek high-reliability parts. *Electronics* (September 1975).

[4] H. Raiffa (1968) *Decision Analysis.* Addison-Wesley, Reading, Massachusetts.

[5] B. Ronen and J. S. Pliskin (1981) Decision analysis in microelectronic reliability: optimal design and packaging of a diode array. *Opns Res.* **29**, 229–242.

[6] *Military Standardization Handbook 217C, Reliability Prediction of Electronic Equipment.* Department of Defense, U.S.A. (September 1980).

[7] *SAM-D Reliability Prediction Notebook.* Assurance Engineering Department, Raytheon Company, Missile Systems Division, Bedford, Massachusetts (June 1973) (BR 7675).

[8] H. Siddsons and E. Sowton (1967) *Cardiac Pacemakers.* Charles C. Thomas, Springfield, Illinois.

3.4   BUDGET PLANNING FOR PRODUCT ENGINEERING

Keefer and Kirkwood's paper from 1978 provides an illustration of the use of multi-attribute utility functions. Note how preferential and utility independence conditions were verified, thus justifying the use of a multiplicative utility function.

In assessing utilities, the questioning was structured in such a way that the confounded problems of identifying attitude to risk and trade-offs between attributes were separated. We have already remarked that this feature of the theory is of central importance in enabling the assessment of multi-attribute utilities. Indeed, Keefer and Kirkwood found it to be so. The decision maker had difficulty in answering questions in which he was simultaneously required to react to uncertainty and to make trade-offs. Fortunately, simpler questions can be and were asked.

Also noteworthy in this study is the fact that a decision tree representation was not used. The decision variables were continuous, and there was neither a time dimension nor any contingent decisions. The analysis was, therefore, structured on the continuous conterpart of a decision table. Expected utilities were calculated as integrals rather than sums.

Finally, notice again the importance of sensitivity analysis in bringing qualitative understanding. It is becoming a clear and recurring theme.

A Multiobjective Decision Analysis: Budget Planning for Product Engineering

DONALD L. KEEFER and CRAIG W. KIRKWOOD

Gulf Management Sciences Group, Gulf Oil Corporation
and The University of Michigan

A multiobjective decision analysis was conducted to help the director of a product engineering group within a major corporation plan the allocation of his operating budget. Use of decomposition results from multiattribute utility theory led to a nonlinear programming formulation of the allocation problem that was convenient to solve and for which the required utility and probability data could be obtained. The analysis indicated that a substantial change from the traditional allocation policy would be desirable. The analysis approach and the results were well received by management.

INTRODUCTION

This paper describes a multiobjective decision analysis conducted with the director of a large product engineering department within a major corporation to help him allocate his annual operating budget. This department has engineering design responsibility for several major product lines involving the same general product type. Engineering effort is concentrated in three areas:

(1) *Cost improvement*: reducing the cost of the product,
(2) *Quality:* preventing and responding to field incidence of product failures, and
(3) *New features and models*: developing new features for existing product models, periodically revising the major model lines, and responding to requests for special limited edition models.

The annual planning problem is to allocate the operating budget among the three areas in order to do "as well as possible" in each. Since departmental resources are limited, tradeoffs must be made among competing objectives. Also, there is substantial uncertainty about the levels of performance that would result from various allocations among the areas. These complications are compounded by the lack of established measures of departmental performance in the quality and new features and models areas. Furthermore, there are conflicting pressures from other groups within the corporation to emphasize the areas in which they have vested interests.

Due to these complexities, the allocation had always been made by informal, intuitive means. However, the department director believed that a more formal analysis might be beneficial. The analysis reported here quantified the tradeoffs and uncertainties in the director's allocation problem and his preferences regarding them. It concentrated on the overall allocation planning problem rather than on the details and timing of specific projects within each area of effort.

The paper discusses previous work related to decision-making with uncertainties and multiple objectives, the analysis approach used here, the data assessment and solution procedures used and the impact of the analysis.[1]

RELATED PREVIOUS WORK

The literature from organization theory shows that uncertainties and multiple conflicting objectives are common in management decision problems, and often are too difficult for the unaided human mind.[2,3] Unfortunately, most quantitative techniques designed to aid managers in making complex decisions do not take these two factors into account simultaneously in a formal manner. For example, Baker and Freeland[4] cite the inadequate treatment of multiple decision criteria and uncertainties as a major limitation of current analysis methods in R & D planning. Lee[5] notes that "a formal decision analysis which is capable of handling multiple conflicting objectives by taking priorities into account may be a new frontier of management technology".

In practice most of these quantitative methods are *ad hoc* approximate procedures[6] whose results lack a clear interpretation. The form of the "scoring" or "utility" functions used to handle the multiple objectives is often postulated for convenience rather than derived from specific verifiable assumptions about the decision-maker's preferences, making it difficult to determine the strengths and weaknesses of the procedures.

Recently a number of studies have proceeded in a more formal manner by using multiattribute utility theory. They verified conditions on the decision-maker's preferences to establish the form of a multiattribute utility function which was subsequently used in the analysis of conflicting objectives and uncertainty. Keeney has applied this approach to planning for airport expansion,[7] and Keeney and Nair[8] have applied it to the siting of nuclear power plants. Bell[9] has used similar methods in the evaluation of policies for the control of a forest pest, while Hax and Wiig[10] have analysed a capital investment problem in the mining industry. Methods similar to these were applied in the budget allocation problem considered in this paper; however, in this application the decision variables were continuous.

ANALYSIS APPROACH

If $x = (x_1, x_2, \ldots, x_K)$ is the vector of budget allocations to each area of departmental effort, then the decision problem is to select the best value of x. To analyse this problem as a multiobjective decision problem using multiattribute utility theory[11] a set of measures of effectiveness, or attributes, $R = (R_1, R_2, \ldots, R_N)$ is defined to indicate the extent to which the different objectives of the decision problem are met by any specific outcome. If the axioms of rational choice,[11-15] which most people find convincing as a basis for normative decision-making, are to be met then the decision-maker should select the value of x which maximizes his expected utility

$$\int_R u(r)f(r|x)\,dr,\tag{1}$$

where $r = (r_1, r_2, \ldots, r_N)$ represents a specific value of R, f is the decision-maker's probability density function over R given the value of x, and u is his utility function.

Determining the feasible budget allocations, the proper attributes, and the functions u and f are now considered.

Feasible budget allocations

Preliminary discussion with the department director indicated that it would be useful to concentrate on the allocation of the yearly departmental operating budget among the three areas of effort for a normal year under stable year-to-year conditions. This assumption of stability had been appropriate during a recent 4-year period, and the director felt it would be an adequate approximation for most future planning situations. Under normal conditions the total departmental budget remained roughly constant in real terms from year to year. Consequently, the director was used to thinking in terms of the fraction of his budget allocated to each area rather than the actual amount. Thus, it was decided to specify a budget allocation by $x = (x_1, x_2, x_3)$, where x_1 is the budget fraction for cost improvement, x_2 is the budget fraction for quality, and x_3 is the budget fraction for new features and models.

Factors such as organizational pressures exerted by various groups that interface with the product engineering department and the skills available within the department limit the director's flexibility in setting allocation policy. Taking these bounds into account, along with the obvious constraint that the planned budget fractions must sum to one, the problem can be stated as

$$\underset{x}{\text{maximize}} \quad \int_R u(r)f(r|x)\,dr \tag{2a}$$

subject to

$$a_k \leqslant x_k \leqslant b_k, \quad k = 1, 2, 3, \tag{2b}$$

and

$$x_1 + x_2 + x_3 = 1, \tag{2c}$$

where a_k, b_k, $k = 1, 2, 3$, are constants.

Attributes

Defining measures of effectiveness (attributes) was a difficult and crucial part of this analysis. In this problem the primary measure of the value of a particular allocation of resources is how it contributes to the profitability (both short and long term) of the corporation. However, the connection between the director's decisions and this ultimate measure is very indirect, particularly in the quality and new features and models areas. The director and others in the organization were accustomed to considering performance in each area of effort separately, and after some discussion it became clear that there should be separate attributes to measure effectiveness in each of the three areas.

For cost improvement the accounting system credits the product engineering department for cost improvements during a 12 month period after the improvement has been incorporated into production. Because of time delays in the system only a relatively small percentage of a given year's cost improvement credit actually results from that year's efforts, the balance being due to previous accomplishments. Hence two attributes related to the information provided by accounting were used in this study:

(1) *Current cost improvement*: this accounts for the cost improvement in a year due to engineering efforts during that year, and
(2) *Carryover cost improvement*: this accounts for cost improvement in future years due to effort in a particular year.

There were no established measures of effectiveness in the quality and new features and models areas. After careful consideration it was decided to use the following 0–4 scales for these attributes:

(i) For quality
0 = marginally acceptable
1 = fair
2 = good
3 = very good
4 = great.
(ii) For new features and models
0 = poor
1 = fair
2 = good
3 = very good
4 = great.

Since these scales were subjective, care was taken to be sure that each numerical value shown had a well-specified meaning. For example, a "good" rating on the new features and models scale meant that there was effective and prompt handling of work required for new models, together with some progress on a new feature of significant interest. Although specific meanings were elicited for only the five values of each scale listed above, the director ascribed meaning to intermediate values as well and used them as the analysis proceeded. Hence the quality and new features and models scales were both treated as continuous.

Further consultation with the director established that the four attributes discussed would be adequate for describing the consequences of an annual budget allocation, and thus it was decided to use the scales

R_1 = current cost improvement in dollars/unit,
R_2 = carryover cost improvement in dollars/unit,
R_3 = rating on quality scale, and
R_4 = rating on new features and models scale.

Probability distribution: performance–spending relationships

Discussions with the director indicated that performance along any of the four attributes essentially depends only on spending in the corresponding area – not on the other achievement levels or spending levels. In other words, each output achievement variable R_n is probabilistically independent of the others, given the value of the corresponding budget fraction $\hat{x}_n$. Clearly this is an approximation since one can envision individual projects that could affect achievement along more than one attribute – e.g. both R_2 and R_3. However, the costs of such projects can be allocated between the areas affected, thereby diminishing any cross-area spending dependencies. Moreover, the focus here is on the aggregate effects of large numbers of projects, and in this context the director believes the assumption is justified.

Thus the joint probability density function appearing in (1) and (2) can be expressed as

$$f(r_1, r_2, r_3, r_4 | x_1, x_2, x_3) = \prod_{n=1}^{4} f_n(r_n | \hat{x}_n), \tag{3a}$$

where

$$\hat{x}_1 = x_1, \quad \hat{x}_2 = x_1, \quad \hat{x}_3 = x_2, \quad \hat{x}_4 = x_3 \tag{3b}$$

for convenience, and $f_n(r_n | \hat{x}_n)$ is the conditional probability density function over R_n given $\hat{x}_n$. The determination of the f_ns is considered later in this paper.

Utility function structure

Careful questioning of the director verified that certain utility and preferential independence conditions held for his preferences. For convenience let $\bar{R}_i = \{R_1, R_2, \ldots, R_{i-1}, R_{i+1}, \ldots, R_N\}$ and $\bar{R}_{ij} = \{R_1, R_2, \ldots, R_{i-1}, R_{i+1} R_{j-1}, R_{j+1}, \ldots, R_N\}$. Then R_i is utility independent of $\bar{R}_i$ if preferences for risky choices (lotteries) over R_i with the value of $\bar{R}_i$ held fixed do not depend on the fixed value of $\bar{R}_i$. The set $\{R_i, R_j\}$ is preferentially independent of $\bar{R}_{ij}$ if preferences for consequences differing only in the value of R_i and R_j do not depend on the value of $\bar{R}_{ij}$.

Keeney[16] shows:

Theorem. Let $R_1, R_2, \ldots, R_N$ be the attributes of a decision problem with $N \geq 3$. If, for some R_i, $\{R_i, R_j\}$ is preferentially independent of $\bar{R}_{ij}$ for all $j \neq i$, and R_i is utility independent of $\bar{R}_i$, then either

$$u(r_1, r_2, \ldots, r_N) = \sum_{n=1}^{N} k_n u_n(r_n) \tag{4a}$$

or

$$1 + Ku(r_1, r_2, \ldots, r_N) = \prod_{n=1}^{N} [1 + Kk_n u_n(r_n)], \tag{4b}$$

where u and the u_ns are utility functions scaled from zero to one, the k_ns are scaling constants with $0 < k_n < 1$, and $K > -1$ is a nonzero scaling constant.

The procedure developed by Keeney was used to verify that the conditions of this theorem held for the budget allocation problem (see refs 1, 7, 11, 16). As an example of the methods used in the verification procedure, consider the question of whether the quality and new features and models attributes R_3 and R_4 are preferentially independent of the other attributes. The department director was asked what value of r_3 would make him indifferent between $(0.20, \$1.00, 1.0, 3.5)$ and $(\$0.20, \$1.00, r_3, 0.5)$ where $r_1 = \$0.20$ and $r_2 = \$1.00$ are relatively low levels of these attributes. He chose $r_3 = 2.0$. Then he was asked the same question for $\$0.70, \$1.80, 1.0, 3.5)$ and $(\$0.70, \$1.80, r_3, 0.5)$ where $r_1 = \$0.70$ and $r_2 = \$1.80$ are fairly high levels of the attributes and again he chose $r_3 = 2.0$. He agreed that r_3 would remain at 2.0 for any values of R_1 and R_2 within the range of interest in this problem, as long as they were identical for the two alternatives, which supports the assumption that $\{R_3, R_4\}$ is preferentially independent of $\{R_1, R_2\}$.

The questions asked in verifying utility independence required the consideration of simple situations with uncertainty. Although it was necessary to specify carefully the situations and make sure the questions being asked were fully understood, the director had little difficulty thinking about them and, after careful thought, responding.

Further questioning using Keeney's procedures showed that the multiplicative form (4b) was appropriate for this problem. Combining (4b) with (2) and (3) resulted in a final formal statement of the decision problem:

$$\underset{x_1, x_2, x_3}{\text{maximize}} \frac{1}{K} \left\{ \sum_{n=1}^{4} [1 + Kk_n \bar{u}_n(\hat{x}_n)] - 1 \right\}, \tag{5a}$$

where

$$\bar{u}_n(\hat{x}_n) = \int_{R_n} u_n(r_n) f_n(r_n | \hat{x}_n) \, dr_n, \tag{5b}$$

subject to the constraints in (2b, c). To complete the analysis, the various u_ns, f_ns and scaling constants had to be assessed.

DATA ASSESSMENT

Probability distributions

Although it might be possible to construct a stochastic process model to generate the probability distributions in (5b), the director felt this effort was not warranted. Therefore, we chose to assess directly the f_ns with him. An examination of (5) shows that it is necessary, in theory, to assess f_n for an infinite number of values of $\hat{x}_n$. However, it appeared that assessing the distribution for the highest, lowest and a middle value of $\hat{x}_n$, and then interpolating these distributions to obtain those for other values of $\hat{x}_n$ would be adequate. Assessing additional distributions would not be useful since the director would obtain them by mentally interpolating between the low, middle and high distributions already assessed and a relatively simple interpolation scheme proved to be adequate.

Thus, 12 probability distributions (three for each attribute), conditional on the associated spending levels, were assessed. At least five points (the 0·01, 0·25, 0·50, 0·75 and 0·99 fractiles) were obtained for each distribution using standard methods.[15,17,18] For example, to assess $f_1(r_1 | \hat{x}_1)$ for a specified $\hat{x}_1$, the director was asked for the value $r_1(0·99)$ below which it was "almost certain" (i.e. had a 0·99 probability) current cost improvement would fall given the specified $\hat{x}_1$. Similarly, $r_1(0·01)$ was assessed as the value which had a 0·99 probability of being exceeded. Then, a value $r_1(0·50)$ was assessed such that the director believed current cost improvement was equally likely to be above or below $r_1(0·50)$. Similar questions were asked to obtain further fractiles. In addition, various consistency checks were made.[1]

Utility functions

The conditional utility functions u_n in (5b) were assessed using the standard lottery technique.[11,15,17] As an example of the questions asked, consider the quality scale R_3. The director was asked to visualize a situation where R_1, R_2 and R_4 were at some fixed values, and he was faced with a 50:50 chance of R_3 being 0 (i.e. marginally acceptable) or 4 (i.e. great). He was asked for the certainty equivalent of R_3, i.e. the lowest value of R_3 that he would take in exchange (along with the previously specified values of R_1, R_2 and R_4) for the lottery (i.e. uncertain situation). The answers to this question and several similar ones allow $u_3(r_3)$ to be constructed.

Consistency checks were used in the questioning procedure. Although minor inconsistencies were found, a sensitivity analysis showed that they were too small to affect the optimal solution.[1]

Scaling constants

The scaling constants in (4b) and (5a) were initially assessed using a lottery technique.[11,16] For example, to assess k_1 the director was asked to consider a situation where R_1 was at the best level possible in the budget allocation problem and the rest of the attributes were at the worst possible levels. He was asked to compare this with a situation where there was a probability p of obtaining all four attributes at their best possible levels and a probability $1 - p$ of obtaining all of them at their worst possible levels. He selected a value for p which would make the two situations equally preferable to him. This value determined k_1.

The department director found this type of question to be the most difficult of those required for the study since it involved multidimensional comparisons over the extremes of the attributes' ranges, as well as uncertainties. Consequently, the values obtained were checked using simpler questions as suggested by Keeney and Raiffa.[11] For instance, the ratio of two scaling constants can be checked by obtaining tradeoff information over two of the attributes under certainty with the other attributes fixed. A sensitivity analysis demonstrated that the relatively small inconsistencies found were not large enough to affect the optimal solution.

SOLUTION AND SENSITIVITY ANALYSIS

Before solving (5), it was necessary to select an interpolation method to determine $\bar{u}_n(\hat{x}_n)$ for $\hat{a}_n \leqslant \hat{x}_n \leqslant \hat{b}_n$, $n = 1, 2, 3, 4$. A number of different methods were studied,[1] and it was concluded that quadratic interpolation over $\bar{u}_n(\hat{x}_n)$ provided adequate accuracy and was relatively efficient computationally. Using the scheme $\bar{u}_n(\hat{x}_n)$ was calculated by numerical integration of (5b) for the three values of $\hat{x}_n$ for which $f_n(r_n|\hat{x}_n)$ had been assessed. Then the quadratic function of $\hat{x}_n$ passing through these three values of $\bar{u}_n(\hat{x}_n)$ was used to calculate $\bar{u}_n(\hat{x}_n)$ for all values of $\hat{x}_n$ between $\hat{a}_n$ and $\hat{b}_n$.[1]

The formulation (5) is a nonlinear programming problem. In this instance, however, it was not necessary to use a sophisticated solution algorithm. The constraint (2c) was used to eliminate x_3, and a simple computer program was then used to calculate and plot values of the objective function (5a) corresponding to points $\{x_1, x_2\}$ over the feasible region. The optimal solution was then obtained by inspection.[1]

When the problem was solved for the data provided by the director, the optimal solution turned out to be $(x_1, x_2, x_3) = (0{\cdot}38, 0{\cdot}32, 0{\cdot}30)$. This was in contrast to the allocation $(0{\cdot}30, 0{\cdot}28, 0{\cdot}42)$ in recent use for a normal year. With the bounds a_k and b_k that had been imposed on the budget fractions, the optimal solution was a vertex of the feasible region and could be interpreted as

(i) spend at the upper limit for quality, and
(ii) divide the rest of the budget by spending at the lower limit for new features and models and allocating as much as possible to cost improvement.

The expected utility for the optimal strategy was $0{\cdot}806$ while that for the existing policy was $0{\cdot}714$. However, these numbers have no physical meaning by themselves. In order to display the differences between the two policies in a manner that was more meaningful to the director, a number of graphical aids were used. The marginal cumulative probability distributions were shown for each of the four attributes for the two policies. Also, the marginal certainty equivalents for each policy along each attribute were displayed. Finally, in an attempt to characterize the differences between the existing and proposed allocation policies by a single physically meaningful number, the difference in utility was projected entirely on to the carryover cost improvement attribute R_2. This was done by finding for each of the two policies certainty equivalents which had the same values of all attributes except R_2. Clearly these certainty equivalents are not unique since many different numbers could be selected for the common values of R_1, R_3, and R_4. For the values selected in this study, the difference in the R_2 values for the two policies was $0·88/unit. Since the volume of the product involved is around two million units per year, this represents a substantial difference. Thus, according to the analysis, a significant improvement in overall effectiveness, as measured using the department director's preferences, could be obtained by shifting to the policy indicated. (Of course, this result does not mean that the indicated policy shift would produce an additional $0·88/unit of carryover cost improvement; the actual performance improvement – the amount of which is uncertain – would be spread over all four attributes.)

Extensive sensitivity studies were carried out to determine the effects on the optimal solution of changes in the u_ns, f_ns and k_ns. Since, with the data provided by the director, the optimal solution lay at a vertex of the feasible region, it was not surprising that moderate changes in the input data did not change the optimal policy. The changes required to alter the optimum policy were considerably larger than the inaccuracies that could reasonably have been present in the data. However, larger changes did move the optimum to other vertices or to points at which none of the bounds (2b) were active.

IMPACT OF THE ANALYSIS

The department director expressed surprise at the optimal policy. Clearly, if he had felt this policy were better than the existing one, he would have been using it instead. He saw this result as a significant contribution to his planning insight. Although he realized there was no way of objectively validating the analysis, he felt that it had gone further in the direction of providing a rational solution to the allocation problem than had been done before.

At the time the analysis was completed (early 1975) he had been pressed, by external business forces (increasing material costs, the economic recession, etc.) to move his policy in the direction indicated by the analysis. Thus, this analysis could not motivate him to change his policy, since the recommended change had already been made. He did indicate that in the light of the analysis he believed he would have changed his allocation policy in the direction indicated even if the changes in external conditions had not occurred. He felt confident that the analysis had captured the essence of the problem: the assumptions were reasonable, and the data were sufficiently accurate to be useful.

When the analysis results were presented, the director requested that the data be revised to represent more accurately the business situation he was currently facing. This was done, and the results were generally the same as with the original analysis – as much as possible should be spent on quality, and as much as possible of the remainder should be spent on cost improvement. This solution reinforced his judgement that his recent shift in budget allocation had been made in the proper direction. He said that this in itself was an important contribution since it is essential that he have confidence in his policies in order to defend them effectively.

The director felt the original "normal year" analysis together with the revised analysis adequately covered the planning situations he could foresee at that point; hence, he did not feel further analysis would be required until the situation changed substantially again.

CONCLUDING REMARKS

As a result of the success of this analysis, a similar one was requested to aid in the allocation planning for the Research and Engineering Division within the same corporation.[1] This study was also successfully completed, and plans have been made to use multiobjective decision analysis as part of the annual budget planning process for that division.

ACKNOWLEDGEMENTS

The authors wish to thank Whirlpool Corporation for financial sponsorship of this research. Special thanks are due to Dr. Gerald Eisenbrandt. This paper was adapted from Donald L. Keefer's Ph.D. dissertation in the Department of Industrial and Operations Engineering, The University of Michigan.[1]

REFERENCES

[1] D. L. Keefer (1976) A decision analysis approach to resource allocation planning problems with multiple objectives. Ph.D. dissertation, Department of Industrial and Operations Engineering, The University of Michigan. Available from University Microfilms. Ann Arbor.

[2] G. T. Allison (1971) *Essence of Decision: Explaining the Cuban Missile Crisis.* Little, Brown, Boston.

[3] R. M. Cyert and J. G. March (1964) *A Behavioral Theory of the Firm.* Prentice-Hall, Englewood Cliffs.

[4] N. Baker and J. Freeland (1975) Recent advances in R & D benefit measurement and project selection methods. *Mgmt Sci.* **21**, 1164–1175.

[5] S. M. Lee (Winter, 1972–1973) Goal programming for decision analysis of multiple objectives. *Sloan Mgmt Rev.* **14**, 11–24.

[6] G. P. Huber (1974) Multi-attribute utility models: a review of field and field-like studies. *Mgmt Sci.* **20**, 1393–1402.

[7] R. L. Keeney (1973) A decision analysis with multiple objectives: the Mexico City airport. *Bell. J. Econ. Mgmt Sci.* **4**, 101–117.

[8] R. L. Keeney and K. Nair (1976) Evaluating potential nuclear power plant sites in the

Pacific Northwest using decision analysis. Professional Paper PP-76-1, International Institute for Applied Systems Analysis, Laxenburg, Austria.

[9] D. E. Bell (1975) A decision analysis of objectives for a forest pest problem. Research Report RR-75-43, International Institute for Applied Systems Analysis, Laxenburg, Austria.

[10] A. C. Hax and K. M. Wiig (Winter, 1976) The use of decision analysis in capital investment problems. *Sloan Mgmt Rev.* **17**, 19–48.

[11] R. L. Keeney and H. Raiffa (1976) *Decisions with Multiple Objectives.* John Wiley, New York.

[12] P. C. Fishburn (1970) *Utility Theory for Decision Making.* John Wiley, New York.

[13] R. A. Howard (1968) The foundations of decision analysis. *IEEE Trans. Sys. Sc. Cyb.* **SSC-4**, 211–219.

[14] J. W. Pratt, H. Raiffa and R. Schlaifer (1964) The foundations of decision under uncertainty. *J. Am. Statist. Ass.* **59**, 353–375.

[15] H. Raiffa (1968) *Decision Analysis.* Addison-Wesley, Reading, MA.

[16] R. L. Keeney (1974) Multiplicative utility functions. *Ops Res.* **22**, 22–34.

[17] R. Schlaifer (1969) *Analysis of Decisions Under Uncertainty.* McGraw-Hill, New York.

[18] C. S. Spetzler and C.-A. S. Stael von Holstein (1975) Probability encoding in decision analysis. *Mgmt Sci.* **22**, 340–358.

3.5   EVALUATION OF NUCLEAR WASTE RISK

The paper of Lathrop and Watson, published in 1982, neither reports an actual case study nor does it describe a potential application of decision analysis *per se*. Rather it discusses the application of a decision-analytic approach to develop a numerical index of the risk inherent in various nuclear waste management policies. I have suggested, and will argue further, that decision analysis provides a framework in which to think about problems. Lathrop and Watson's work is, I believe, an admirable illustration of this.

The development of a hierarchical description of the consequences is of particular note. Here is an example of very complex consequences being summarized usefully in terms of several, but not an overwhelming number of attributes. One may argue that there are other attributes that Lathrop and Watson should have included; but it would be difficult to argue that they have not successfully described the consequences in most important respects.

One problem, which they address and which we avoided, is how to construct utility analyses for groups of people. This problem is fraught with difficulty, not the least being that any concept of group utility may be ill defined (French, 1985, 1986; Keeney and Raiffa, 1976; Raiffa, 1968). Their approach may not be perfect, but it is most certainly worthy of consideration.

Decision Analysis for the Evaluation of Risk in Nuclear Waste Management

JOHN W. LATHROP AND STEPHEN R. WATSON

Woodward-Clyde Consultants, San Francisco, U.S.A. and Emmanuel College, University of Cambridge, U.K.

The implementation of a nuclear waste management technology raises several issues concerning the regulation of social risk. The most basic of those issues are how to regulate a technology when the uncertainties in social consequences are important, and how to incorporate the relevant social values in the regulations. This paper presents a decision analytic approach to resolving these issues, based on the development of radiological risk evaluation indices. While it is essentially a case study, describing work carried out for the U.S. Nuclear Regulatory Commission, this case is used to discuss the more general issues involved.

We begin by discussing the need for risk evaluation to provide a clear and defensible basis for regulating technologies involving social risk. We then present a development of risk evaluation indices for the regulation of nuclear waste management. The indices developed are expected utilities, based on preferences elicited from groups of people. The use of the indices developed is illustrated in a hypothetical example, and the usefulness of the methodology evaluated.

INTRODUCTION

Over the past two decades or so there has been increasing concern over the use of many modern technologies. While much of this concern results from the demonstrable and existing deleterious side-effects of these technologies (for example, the effect on animal and plant life of pollution from many chemical plants, or the cumulative effects in the food chain of insecticides like DDT), there are some technologies for which the cause for concern is potential rather than actual (for example the depletion of the ozone layer by chlorofluorocarbons and by the exhaust gases of supersonic transports, or potential radiation releases for buried nuclear wastes). Governments are concerned to control both kinds of technological side-effects, but the appropriate regulatory action is much more difficult to determine in the latter case, where there is considerable uncertainty about the possible nature of the consequences of the technology and where they can occur in the distant future.

In this paper we discuss how the paradigm of decision analysis might be used to establish an index which can evaluate the risk associated with activities of the latter kind; in particular we shall exemplify our suggestions throughout by reference to the problem of regulating nuclear waste management. At the outset we ought to stress that we are not using 'risk' in the sense normally used in decision analysis. In bringing together decision analysis and risk evaluation, we must adopt terms that clearly distinguish concepts often confused between the two fields. In particular, 'risk' will be used to describe the potential for deleterious consequences associated with a technology, while 'uncertainty' will be used to describe the lack of information available concerning what the impacts of a technology might be.

The need for risks to be analyzed if they are to be adequately managed is beginning to be widely recognized. The important studies by Rowe[1] and Lowrance[2] on the nature and management of risk have been followed by a collection of papers outlining research needs and opportunities in this area, edited by Kates.[3] Our goal in this paper is to add to the emerging literature of this discipline by describing a particular approach to risk assessment, using decision analysis. This is a controversial area; for example, the criticisms by Lovins[4] and Cochran[5] of the decision analysis of alternatives for electricity generation carried out by Barrager et al.,[6] are paralleled by Hoos' attack[7] on the use of risk analysis for nuclear waste management and the critical comments of Tocher[8] concerning the use of decision analytic concepts in any social planning activity. These authors give many detailed arguments against particular quantitative methods; the reader is urged to consult them as an antidote to overconfidence in the methods of systems analysis, and decision analysis in particular. However, it is fair to say that most of their criticisms are destructive in nature and often might easily be paralleled by similar criticisms of alternative informal and traditional methods of decision making and regulation setting. It is our strong belief that if used properly, the quantitative methods of decision analysis, while subject to some valid criticism, improve decision making in that they provide a consistent base for analysis and improve communication. In many cases the informal methods share all the flaws of the formal approach; it is just that their very informality obscures the fact that these flaws exist.

In the next section we describe in detail a decision analytic approach to the evaluation of risk and how this method was used to construct risk evaluation indices for the regulation of nuclear waste management. The following section describes the results of our work in terms of a hypothetical example, which also serves to illustrate the role of the approach in the development of regulations. The last section briefly comments on the present and potential usefulness of the approach. This work was carried out for the U.S. Nuclear Regulatory Commission (N.R.C.) under the direction of Lawrence Livermore National Laboratory. Larger reports (Lathrop,[9] Watson and Campbell[10]) contain more detailed descriptions of the project and the results of the study.

DEVELOPMENT OF A RISK EVALUATION INDEX

Quantitative approaches to risk measurement have quite a long history. (See Farmer,[11] Starr[12] and the bibliography of Clark and Van Horn.[13]) Broadly speaking, they have all recognized that a measure of risk should be an increasing function of the probabilities of deleterious consequences and the severity of those consequences. A measure often chosen has been expected fatalities. Shortcomings of this criterion include its inability to cope with attitudes toward uncertainty

(Who feels that a 50–50 chance of two deaths is just as bad as one certain fatality?), and the exclusion from consideration of consequences other than death (which in the case of radiation exposure includes effects as significant as genetic mutations). Papp *et al.*[14] recognized this inadequacy and suggested that it should be rectified by using utility theory; the present paper describes how we have followed this suggestion for nuclear waste management.

Very briefly put, our approach consists of developing a multidimensional utility function over the health-effect consequences of a nuclear waste management system, separately assessing a probability distribution over those consequences for each alternative system, and then calculating the risk index for each system as its expected utility. Because our utility function increases with the severity of negative consequences (contrary to convention), the risk index is an increasing function of probability and severity of consequences, as desired. As explained by Howard,[15] the expected utility represents preferences for uncertain outcomes on a cardinal scale, ranking complex alternatives in a manner consistent with preferences revealed in comparisons of simple alternatives. While the basic ideas of this approach are relatively straightforward, there are several steps in its implementation. We shall now describe those steps as we encountered them in our development of a risk index, some of the problems each step entailed and how we dealt with them.

Consequence scope

In constructing a utility function describing consequences, we first had to specify the important possible consequences of a nuclear waste repository. The most significant possible effects are, clearly, health effects on humans, but there are many others, such as effects on animal and plant life, restrictions of civil liberties and restrictions on land use. We made a modelling decision at this stage to limit attention to health effects on humans. This limits, of course, the use to which our risk index can be put, since in comparing repositories any of the other possible consequences may be of significance. However, the purpose of this study was not to analyze specific decisions on the location and design of repositories, let alone whether or not radioactive waste should be generated; it had the lesser, but still important goal of providing a measuring device for only one aspect of the waste management problem, namely the health risk to humans. Here is an example of the methods of decision analysis contributing to public policy formation by analyzing part of the problem. In our experience, such partial analyses are much more likely to be used and useful than any attempt to bring the whole decision making process for an issue of public policy under the hammer of hard analytic methods.

Value source

The idea of using decision analysis on public policy issues is, of course, not new. Both Howard[15] and Edwards[16] have proposed schemes for social decision analysis. One of the problems which such schemes raise, however, is the determination of whose values should be represented in the utility function. In an evaluation of the risk in waste management it would seem desirable somehow to reflect public values; but these may differ importantly from person to person – how may such differences be combined into one index? It should be helpful here to find some method of aggregating individual utility functions to create a group utility function. While there are difficulties in specifying such a procedure (as Seaver,[17] p. 14, observes: "no entirely satisfactory method for devising group utilities exists"), the work of Keeney and Kirkwood[18] cites theoretical support for the idea of using a weighted sum of utility functions as a group utility function, the weights to be determined by a "benevolent dictator" or an "honest broker". At the same time, the weights given to people with markedly different stands on a policy issue represent trade-offs perhaps best made in the course of the political process, not in the risk analysis.

The relevant political process in this case centres on the N.R.C., which is effectively charged with balancing the political interests of any conflicting parties. There is always reason to suspect that any regulatory agency is subject to disproportionate pressure from the industry it is supposed to monitor, so that such as agency may not be an ideal umpire of conflicting interests (see, e.g. Hoos[7]). However, the goal of the approach described here is not to reform the political process, but simply to develop a risk evaluation methodology that incorporates social values into the existing political process. Consequently, we drew the bounds between the evaluation and the process it is to serve and decided to represent separately the values for four groups, described below. The groups chosen were ones to which the N.R.C. is responsive, which span the political spectrum of interests faced by the N.R.C., and within which values could be expected to be relatively homogeneous. While this strategy avoids aggregating values across conflicting groups of people, values must still be aggregated across individuals within each homogeneous group to arrive at a group utility function. That aggregation is explained later.

In the study, 58 people were interviewed and utility functions elicited from each of them. They were divided into groups as follows:

(a) *National Advisors* (13 respondents). This group consisted of persons who had the ear of the Federal policy-makers in that they either served on nuclear waste advisory bodies, or their views were published or otherwise consulted by Government policy-makers.

(b) *Concerned Citizens* (33 respondents). The intent in this case was to select citizens who were at least somewhat abreast of social issues and who were concerned that Government actions should reflect the general public interest.

(c) *Nuclear Power Opponents* (7 respondents). These were persons who were known publicly to oppose further development of nuclear power, at least until safety problems have been resolved.

(d) *Nuclear Power Advocates* (5 respondents). These were persons who had been identified as advocates of the further development of nuclear power. Some of the respondents in this group maintained that they did not consider themselves as advocates of nuclear power.

Group utility functions for each of these four groups were calculated as weighted sums of the individual utility functions. Then these were combined with example probability distributions describing different repositories, producing four distinct risk evaluation indices for each repository. We return to the results of these calculations later. Because of our decision to elicit utility functions from so many people, it was necessary to construct a simple standard form for the utility functions, which called for several approximations in the following steps.

Consequences

One of the basic problems of approximation concerned how to describe health effects. The first widely used and definitive discussion on the health effects of radiation is the 1972 report of the U.S. National Academy of Sciences Advisory Committee on the Biological Effects of Ionizing Radiation[19] (the BEIR report). It is apparent from this and other documents (see ICRP No. 26[20]) that the health consequences of radiation exposure are many and varied. Despite this, there is a fairly obvious categorization, as portrayed in Figure 1. The first distinction is between "stochastic" and "non-stochastic" effects. To quote ICRP No. 26[20], p. 2:

"Stochastic effects are those for which the probability of an effect occurring, rather than its severity, is regarded as a function of dose, without threshold. Non-stochastic effects are those for which the severity of the effect varies with the dose, and for which a threshold may therefore occur."

Stochastic effects can be further categorized as either somatic, if they become manifest in the exposed individual himself, or genetic, if they affect his descendants. It is clear that within each category there is a very large number of possible effects. The modelling decision that now faced us was how many of these to include in a list of attributes for a utility function. Here a balance had to be drawn, as always in applied decision analysis, between analytic simplicity and completeness. We needed to find a set of attributes which included everything important in the evaluation of risk, and yet was small enough for a utility function over the attributes to be elicited reasonably easily from a large number of people in a short time. We selected the following five variables, as indicated in Figure 1:

x_1 – number of fatal cancers;
x_2 – number of non-fatal cancers;
x_3 – number of mutations;
x_4 – number of acute fatalities;
x_5 – number of cases of impaired fertility.

Acute fatalities here refers to those that occur shortly after exposure, and so correspond to incidents of very high dose associated with pre-seal accidents.

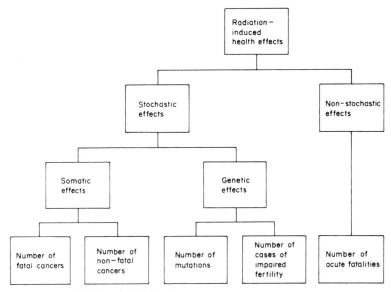

Figure 1 Categorization of health effects.

Several approximations and assumptions about the respondents' value structures are implied by the above list. First, the use of the total number of cases of a particular kind implies that any such case is as bad as any other such case. Second, there are other known radiation effects which do not appear on the list, such as cataracts or retarded development in children. Third, some of these categories cover a very wide class, and in order to enable trade-offs between such effects to be accessible psychologically to our respondents, it was necessary to particularize them considerably; thus the class of mutations was represented by a mentally subnormal person who needs some extra care throughout his life. This last approximation will only be good if this case is an approximate certainty equivalent of the class of all possible mutations; this in turn depends upon the nature of the probability distributions with which the utility function will be used, and we had to complete this study without a good idea of those probabilities.

We are not satisfied that this list of attributes is the best set to describe health effects, and we feel that more work should be done on determining a good set. However, this seems to us a reasonable first attempt and does satisfy some of the criteria for choosing attributes: sufficient complexity to cover the important structure of the problem, yet sufficient simplicity for elicitations of the trade-offs to be believable and available in a reasonably short period of time.

The next point concerns the circumstances in which the radiation doses arise. As Fischhoff et al.[21] discovered, the circumstances of risks do affect their importance

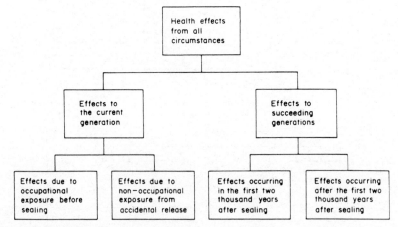

Figure 2 Categorization of the circumstances of radiation exposure.

as judged by members of the public. Of the many possible circumstances, the two we judged to be significant enough to include explicitly in the utility function were whether or not the risk was undertaken as part of a person's occupation and the time at which the radiation dose might arise. Figure 2 illustrates this categorization. Notice that occupational risk can only be suffered by workers in the current generation, since after sealing no further operation of the repository will be necessary.

We define an index i to indicate the circumstances of the dose giving rise to a particular health effect according to the following scheme:

$i = 1$: effects due to occupational exposure before sealing;

$i = 2$: effects due to non-occupational exposure, to the current generation;

$i = 3$: effects caused by a dose in the two thousand years following repository sealing;

$i = 4$: effects caused by a dose more than two thousand years following repository sealing.

Each of the five health effects can occur as a result of circumstances in each of the four categories above. There are, therefore, twenty variables x_{ij} which describe the possible effects caused by a nuclear waste repository, where $i = 1, \ldots, 4$ indicates the cause of the dose and $j = 1, \ldots, 5$ indexes the type of health effects (e.g. x_{32} is the total number of non-fatal cancers arising from doses received from the nuclear waste in the two thousand years after the repository has been sealed). Note that the time division here is somewhat coarse. Once again, a balance had to be struck between capturing the essential elements of the repondents' value structures and keeping the model reasonably simple.

Individual utility functions

The next problem was to establish a parametric structure for the utility function of each individual. We chose to construct first a value function $v(\underline{x})$ (where $\underline{x}$ is the vector whose twenty components are x_{ij}), which would have the property that any possible set of health effects x judged to be equivalent to or worse than any other y would have $v(x) \geqslant v(y)$. The form we adopted was

$$v(\underline{x}) = \sum_{i=1}^{4} \sum_{j=1}^{5} \alpha_i \beta_j x_{ij}.$$

This form makes some rather sweeping assumptions about preference structures. First, the linearity implies that, for example, if one extra acute fatality is as bad as two extra mutations at one level of all the variables, the same is true at all other levels of the variables. Second, the fact that the constant coefficient of each variable x_{ij} is a product $\alpha_i \beta_j$ implies that trade-offs among health effects do not depend on the circumstances of the dose, and trade-offs between the same effects in different circumstances do not depend on which health effect is considered. Although these assumptions are fairly strong, our respondents' preference structures seemed to be consistent with them to an adequate approximation.

The final stage in the construction of a von Neumann–Morgenstern utility function for each of our respondents was to reflect uncertainty preference (our synonym for the more usual 'risk preference', which we shall avoid to prevent confusion with the more general use of the word 'risk' in this paper). We constructed a single-argument utility function reflecting uncertainty preference, whose argument was $v(\underline{x})$. In our detailed reports[9,10] we used an exponential family of utility functions, which has the advantage of being a single-parameter family. This proved to be an inadequately rich family to describe the uncertainty preferences of many of our respondents, at least as they were assessed in this study. In a re-analysis of our work, a two-parameter family of utility functions would be desirable.

Note that this utility function incorporates a very straightforward solution to a central problem with risk evaluation of nuclear waste management: how much relative weight to give health effects to future generations as opposed to health effects to the current generation. This is an exceedingly important question, which in many analyses is answered by the choice of a discount rate. Many authors have criticized this approach; a good account is given by Goodin.[22] He points out that the economic opportunity-cost arguments for discounting do not readily apply to the long time periods involved in nuclear waste management, since an appropriate rate of return on resources saved is extremely difficult to estimate, even in the form of a subjective probability distribution. Goodin goes on to describe and attack various other rationales for discounting. The strength of our approach is that it does not presume any particular prescriptive rationale. It simply asks respondents, in a trade-off format explained below, for their relative weights between current and future health effects. A respondent is free to take a stand that future health effects should be weighted equally to current ones, or to ascribe to a discounting rationale and set his or her own rate of discount (positive or negative). If there were a clear consensus from a broad community of experts as to some correct, non-intuitive approach to relative evaluations of health effects over time, our value elicitation could be attacked. But clearly, as Goodin's article attests, there is no such consensus, so it could be argued that our respondents' answers to this particular trade-off question represent as good a basis for setting policy as conflicting expert opinions.

Value elicitation

Having developed the structure of the utility function, the values of the parameters had to be elicited from each individual respondent. One of our team sought out and interviewed 58 respondents (as described above) and made estimates of the parameters $\{\alpha\}$ and $\{\beta\}$ for each person interviewed. He did this by first asking the respondent to order the five health effects as to seriousness, and then making pairwise comparisons. For example, if the respondent stated that a fatal cancer was worse than an acute fatality, he was asked how many acute fatalities was as bad as one fatal cancer. By such comparisons and associated cross-checks the parameters were estimated. Finally, lottery type questions were asked of the respondents, from which their uncertainty preference was assessed.

Because of time limitations it was only possible to spend an hour or so with each respondent. Given the known effects of expressed value changing over time and depending on the type of questions asked (the problem of labile values, Fischhoff *et al.*[23]), we have considerable doubts as to whether our results would have been the same if a longer time had been taken in eliciting values.

Value aggregation within groups

The next step in the calculation of risk indices for each of the four groups mentioned above was the determination of the weights to apply to the utility function of each individual within each group to obtain the group utility function as a weighted sum. As mentioned before, according to Keeney and Kirkwood,[18] and Keeney and Raiffa,[24] p. 539, one reasonable approach to determine the weights in an additive group utility function is to use the judgment of a "benevolent dictator" or "supra decision-maker", who makes a fair balance between the intensities of preference of the group members. This would seem a particularly appropriate approach in this case, since the U.S. Nuclear Regulatory Commission has the responsibility of regulating the nuclear industry in the U.S. in the balance of the public interest. One could ask the Commissioners themselves to determine relative weights between individuals within each group. The nature of our study, however, precluded us approaching the Commissioners. We had to resort, therefore, to a procedure for determining weights based on the concept of equal weighting of calibrated utility functions.

The process of calibrating individual utility functions consists of weighting each function so that a utile on one is in some sense comparable to or commensurate with a utile on another. Even this limited operation involves interpersonal comparisons of utility and so has no fully satisfactory methodology. Our approach was to standardize the utility functions on a unit interval: one fatal cancer caused by a pre-seal non-occupational low dose. This standard was chosen because it was the most clearly understood health effect, so that the effect of differing individual interpretations of a health effect on calibration error was minimized. In addition, this standard involved the dose circumstance generally used as a basis by the respondents against which other dose circumstances were discounted, so that the effect of differing discounts on calibration error was minimized.

Probability distributions

Now that we had constructed four utility functions over the twenty-component health effects vector x, each derived from different groups of people, we could combine these utility functions with probability distributions over x describing the uncertainty in the consequences from any particular repository. Here we faced a further problem. Although the theory of subjective probability implies that probability distributions can be constructed describing any degree of uncertainty, the implication of the methods of social decision analysis suggested by Howard[15] and Edwards[16] is that the best available expertise should be harnessed to construct probability distributions on the outcomes of alternative repositories. Lawrence Livermore National Laboratory was at the time of this study engaged in an extensive project to produce the necessary expert knowledge of the likely consequences of any particular repository site and design, but the appropriate results were not available at the time our reports[9,10] were completed. Instead, in order to test the implications of our utility functions, we elicited example subjective probability distributions from two staff members at Lawrence Livermore National Laboratory who had considerable personal experience in the assessment of radiation hazards, and in the possible consequences of geological disposal of radioactive waste. They produced probability distributions for two hypothetical repositories, differing from each other in a way that was meant to represent the effects of a hypothetical regulation requiring an increased amount of waste packaging. It was hypothesized that the regulation would increase the probability of pre-seal health effects slightly (due to increased handling) and decrease the probability of post-seal health effects (due to increased isolation). It should be emphasized that these probability distributions are for example purposes only and are in no way based on the results of the physical repository modelling effort conducted at Lawrence Livermore National Laboratory.

RESULTS

The results of our work fall into two categories: features of the elicited values and characteristics of the risk evaluation index itself, as generated from the assessed preferences of the four respondent groups and applied to the example probability distributions mentioned above.

Elicited values

The two most interesting features of the elicited values happen to coincide with the two main improvements of expected utility over expected fatalities as a risk measure: the representation of attitudes toward uncertainty and the evaluation of more dimensions of health impacts than fatalities alone. Concerning attitudes toward uncertainty, half of the respondents were uncertainty preferring (i.e. risk seeking) in that they preferred a 1% chance of 100 fatal cancers over one fatal cancer for certain. About one quarter of the respondents were uncertainty averse, one quarter uncertainty neutral. In every one of the panels, taken separately, less than half of the panel members were uncertainty averse. These preferences indicate that a group utility function that assumes individual uncertainty attitudes to be neutral would match or overstate the aversion to uncertainty of about three-quarters of the respondents. Because the risk evaluation index is intended to measure the risk of repositories relative to the risk of standards involving less uncertainty, and because any errors in the estimates of repository impacts are apt to understate their uncertainty (see Goodin[22]), and index based on a group utility function that overstates aversion to uncertainty could be considered desirably conservative. For these reasons, and because of several advantages for implementation of an uncertainty neutral group utility function, that form of function was adopted for the risk index calculations performed in this study.

Concerning evaluation of health impacts other than fatalities, value trade-offs between fatal and non-fatal health effects were such that non-fatal health effects contributed significantly to the risk evaluation index. As one example, more than half of the respondents considered a mutation as worse than or equivalent to a fatal cancer.

Characteristics of the index

The clearest way to present the risk index, and the implications for that index of the elicited values, is to demonstrate its use. The following paragraphs will step through a few simple example calculations to that end. The calculations are oriented toward answering three basic questions: does the proposed regulation decrease overall risk, by how much is the overall risk decreased, and does the regulation decrease the risk to an acceptable level?

The most elemental question these risk evaluation indices can help to answer is: does the proposed regulation decrease overall risk? For a regulation requiring increased packaging of wastes, for example, the answer is not immediately clear, since it would decrease post-seal risk at some expense in increased pre-seal risk. The risk indices (expected utilities) calculated according to the scheme outlined in the previous section are given in Table 1, for repositories with and without the regulation, for each of the four respondent groups. Since we adopted an

Table 1. Expected utilities of four respondent groups for two repositories

Repository	National advisors	Concerned citizens	Nuclear opponents	Nuclear advocates
Repository A, without new regulation	2·4	9·1	5·2	0·46
Repository B, with new regulation	2·0	7·3	4·2	0·41

uncertainty neutral (linear) form for the group utility functions, and because of the particular normalizations we used, the numbers in Table 1 are not only expected utilities; they are also equivalent pre-seal non-occupational fatal cancers. As the column differences in Table 1 make clear, the example regulation does in fact increase overall safety for each respondent group. Differences in index value between groups can be explained in terms of preference differences in discounting of future health effects and in the relative importance of different health effects. However, there is no direct operational significance of the differences within rows in Table 1. All that matters as far as regulation selection is concerned are the differences within columns.

There is one basic concept, perhaps not stressed enough in this report, that Table 1 helps make clear. The differences in numbers within either row of the Table should not be interpreted as measurement error. We have not constructed a single risk evaluation index, but a set of four such indices, each capturing, in some way the attitudes of one of the four panels. The risk evaluation index of a repository is not some physical attribute of that repository; rather, it is an evaluation of the impacts of that repository, as judged from a particular set of personal values.

The second basic question the risk evaluation index can help to answer is: how effective is the proposed regulation? Table 2 presents three different ways to scale the difference in risk evaluated between repository A (without the regulation) and

Table 2. Measures of risk reduction due to the proposed regulation

Measure of risk reduction, difference in equivalent:	National advisors	Concerned citizens	Nuclear opponents	Nuclear advocates
...pre-seal non-occupational fatal cancers	0·44	1·8	0·96	0·05
...occupational dose (man-rem)	12,000	34,000	14,000	3000
...pre-seal non-occupational dose (man-rem)	5300	19,000	8800	800

repository B (with the regulation). These differences are measures of the effectiveness of the proposed regulation. The first row of Table 2 is simply the set of differences between rows from Table 1: the reduction of risk in utiles caused by the regulation. The second and third rows of Table 2 can be useful for comparing the effectiveness of a regulation with other regulations or technical alternatives on convenient dimensions. For example, if an alternative regulation to the one used in the example would have the sole effect of reducing occupational dose, Row 2 would offer a very direct comparative measure.

The effectiveness of the regulation could be measured in terms of reduction in expected fatalities, a very different measure from a reduction in equivalent fatal cancers. For the hypothetical probability distributions used in this example, the reduction in expected fatalities due to the regulation comes to 0·23 expected lives saved. It is interesting to compare this figure with the first row of Table 2. For every group except Nuclear Advocates, the fact that the reduction in equivalent fatal cancers incorporates non-fatal health effects leads to measures of regulation effectiveness more than twice as large as the less comprehensive 'expected lives saved' measure. On the other hand, the fact that the Table 2–Row 1 measure incorporates the Nuclear Advocates' discount factors for future and occupational fatalities leads to a measure of regulation effectiveness much smaller than the undiscounted 'expected lives saved' measure. These examples should make clear that the 'expected lives saved' measure is not at all 'value-free'. It in fact makes specific assumptions concerning value trade-offs that seem to be importantly at variance with the value trade-offs assessed from our respondent groups.

The third basic question the risk evaluation index can help to answer is: does the regulation reduce the risk of the repository to an acceptable level? The definition of acceptable risk is quite involved and will not be addressed here (see Lathrop[9] and Watson and Campbell[10]). But whatever the definition of acceptable risk, the risk evaluation index can play a key role in its determination by providing a common scale on which to compare risks and on which to set an acceptable risk limit. For example, suppose some analysis finds that a repository is acceptable if its risk is less than the risk due to a particular occupational population dose. For any given risk evaluation index, this dose can be expressed in utiles as a limit on that index scale. Any repository can be evaluated using the risk evaluation index on the same utile scale, and so directly compared with the acceptable risk limit.

DISCUSSION

The purpose of this paper has been to describe an attempt to use the techniques of decision analysis to establish risk indices for nuclear waste repositories which reflect public values. The novel aspect of our approach, which we have not noticed reported elsewhere, has been the creation of utility functions representing segments of public opinion by first constructing utility functions for a sizeable number of individual respondents and then combining them to form group utility functions. One problem of this approach was that difficulties in eliciting values from so many respondents required us to make rather more approximations in the methodology than would normally be necessary in a study of this kind.

The two primary elicitation results call for some discussion. First, the small fraction of uncertainty averse respondents suggests that if a risk evaluation index is to reflect the popularly observed aversion to catastrophe, it must represent that feature in some other way than the uncertainty aversion of utility theory. Second, the large weight given to non-fatal health effects in our study demonstrates the importance of a risk evaluation index comprehensive enough to include them.

Several interesting results came out of this research and are presented above. However, the most valuable results are the two very general aspects of risk evaluation that our approach has made clear by example. First, risk is not some physical quality of a physical system; it is a function of both the physical system and the group of people evaluating the risk. Second, there is no value-free measure of risk; even the commonly used expected fatalities measure assumes particular values.

In light of the further development required in our methodology as mentioned in our description, we must pose the question: can we recommend that the N.R.C. use the risk evaluation methodology described here to form a basis for the regulation of nuclear waste management, or are the problems so great that this whole approach to risk evaluation should be abandoned? Our answer is a positive recommendation to apply our methodology, but only after further development. It is clear, because of the difficulty of the problems outlined above, that considerable further work needs to be done before we can be confident that the risk evaluation indices created adequately represent the relevant social values and provide a satisfactory means for incorporating those values into the risk management process. As challenging as those problems are, the fact remains that risk must be evaluated before it can be managed, and we feel that the risk evaluation methodology described here is an improvement over the less explicit process of risk evaluation that could probably be used in its stead. While the value elicitation questions called for in our approach were very difficult to answer, the fact remains that those questions must be answered in the course of nuclear waste risk management, either explicitly with a methodology such as the one described here, or implicitly without any formal analysis. All our methodology does is force people to confront these difficult trade-offs directly, rather than leave them to be determined implicitly by a process that manages risk without defining it.

Note that the methodology presented here is not intended to depoliticize what is clearly a political process. The determination of an acceptable level of risk is left entirely to the political domain. All that is suggested here is a measuring rod, so that the political debate surrounding the regulation of social risk is clarified. Even more than that, the concepts of risk proposed here are not normative in nature but are based on social values elicited from groups of people to which the regulatory agency is normally responsive. As several people have pointed out (see, for example, Otway et al.[25]) other issues, such as centralization of power, may be at stake in political debates ostensibly concerned with technological risk. But if risk is more clearly defined, then those other issues will be brought more clearly into focus, instead of being obscured in ill-defined notions of social risk.

On the basic argument that an improvement in the present level of information about public values concerning risk is necessary for the proper regulation of nuclear waste management, we maintain that our approach is worth pursuing. At the very least, it will provide a rational basis for proposed regulations put forward by the staff of the N.R.C. at the start of the long chain of review. But beyond that, we believe that as society presses more toward democratic involvement in the regulatory process, the methodology presented here will become more and more attractive as a fair and just means to reflect social values in regulatory decision making.

ACKNOWLEDGEMENTS

The work on which this paper is based was undertaken while Dr Watson was on leave of absence with Decisions and Designs, Inc., McLean, Virginia, and Dr Lathrop was with the Lawrence Livermore National Laboratory, California, and was supported by subcontract 969 3603 from that laboratory. The authors wish to acknowledge the valuable assistance of Vince Campbell of Decisions and Designs, Inc. Dr Campbell performed the value elicitations described in this paper and made several helpful comments concerning earlier drafts of the paper. The authors also wish to acknowledge several helpful criticisms by referees on an earlier draft.

REFERENCES

[1] W. D. Rowe (1977) An Anatomy of Risk. Wiley, New York.
[2] W. W. Lowrance (1976) Of Acceptable Risk: Science and the Determination of Safety. W. Kaufmann, California.
[3] R. W. Kates (Ed.) (1977) Managing technological hazard: research needs and opportunities. Monograph 25. Institute of Behavioral Science, University of Colorado.
[4] A. B. Lovins (1976) pp. 104–115 of Ref. (6).
[5] T. B. Cochran (1976) pp. 93–94 of Ref. (6).
[6] S. M. Barrager, B. R. Judd and D. W. North (1976) The economic and social costs of coal and nuclear electric generation. NSF 76-502. National Science Foundation, Washington.
[7] I. R. Hoos (1978) Assessment of methodologies for radioactive waste management. In Essays on Issues Relevant to the Regulation of Radioactive Waste Management (W. P. Bishop et al., Eds), pp. 31–46. NUR-EG-0412, U.S. Nuclear Regulatory Commission, Washington.
[8] K. D. Tocher (1977) Planning systems. Phil. Trans. R. Soc. A287, 425–441.
[9] J. W. Lathrop (Ed.) (1978) Development of radiological performance objectives, interim results: trade-offs in attitudes towards radioactive waste. UCID-17925. Lawrence Livermore National Laboratory, California.

[10] S. R. Watson and V. N. Campbell (1978) Radiological performance objectives for radioactive waste derived from public values. Final Report PR78-10-80, Decisions and Designs, Inc., Virginia.

[11] F. R. Farmer (1967) Reactor safety and siting: a proposed risk criterion. *Nucl. Saf.* **8**, 539–548.

[12] C. Starr (1969) Social benefit versus technological risk. *Science* **165**, 1232–1238.

[13] E. M. Clark and A. J. Van Horn (1976) Risk-benefit analysis and public policy: a bibliography. Brookhaven National Laboratory, New York.

[14] R. Papp, P. E. McGrath, L. D. Maxim and F. X. Cook Jr (1974) A new concept in risk analysis for nuclear facilities. *Nuclear News*, pp. 62–65, (November).

[15] R. A. Howard (1975) Social decision analysis, *Proc. IEEE* **63**, 359–371.

[16] W. Edwards (1977) How to use multi-attribute utility measurement for social decision-making. *IEEE Trans. Sys. Man Cybern.* **SMC-7**, 326–339.

[17] D. A. Seaver (1976) Assessment of group preferences and group uncertainty for decision making. Technical Report 001597-4-T, Social Science Research Institute, University of Southern California.

[18] R. L. Keeney and C. W. Kirkwood (1975) Group decision making using cardinal social welfare functions. *Mgmt Sci.* **22**, 430–437.

[19] National Academy of Sciences Advisory Committee on the Biological Effects of Ionizing Radiation (1972) The effects on populations of exposure to low levels of ionizing radiation (known as the BEIR Report). National Research Council, Washington.

[20] International Commission on Radiological Protection (1977) *Annals of the ICRP No. 26*. Pergamon Press, Oxford.

[21] B. Fischhoff, P. Slovic, S. Lichtenstein, S. Read and B. Combs (1978) How safe is safe enough? A psychometric study of attitudes towards technological risks and benefits. *Policy Sci.* **8**, 127–152.

[22] R. E. Goodin (1978) Uncertainty as an excuse for cheating our children: the case of nuclear wastes. *Policy Sci.* **10**, 25–43.

[23] B. Fischhoff, P. Slovic and S. Lichtenstein (1980) Knowing what you want: measuring labile values. In *Cognitive Processes in Choice and Decision Behavior* (T. Wallsten, Ed.). Lawerence Erlbaum, New Jersey.

[24] R. L. Keeney and H. Raiffa (1976) *Decisions With Multiple Objectives: Preferences and Value Trade-offs*. Wiley, New York.

[25] H. J. Otway, D. Maurer and K. Thomas (1978) Nuclear power: the question of public acceptance. *Futures*, pp. 109–118.

3.6   REQUISITE DECISION MODELLING

My understanding of decision analysis owes much to discussions with Larry Phillips. He and his co-workers, first at Brunel University, then at the London School of Economics, have led the growth of successful applications of decision analysis in the UK over the last decade. Moreover, he has been more successful than most in articulating that decision analysis provides a framework for thought and a language in which groups may communicate. In the paper reproduced here, published in 1982, he works through these ideas in the context of a simple case study. It is an excellent paper: even if you skip-read everything else in these notes, pause and read this paper carefully...at least twice.

There are several points that merit close attention in this paper. Our recurring themes are there: the prime output of the analysis was qualitative understanding; and sensitivity analysis had a central role in achieving this. However, the main thrust of the paper is not one of these recurring themes: it is an exploration of the form of modelling used in decision analyses. Remember that in section 1.3 I sounded a warning that modelling within a decision analysis was a distinctly different activity to that in scientific modelling. Larry Phillips explains decision-analytic modelling roughly as follows.

The decision makers begin the analysis ill at ease, discomforted by some half-perceived choice before them. As the analysis proceeds, their perceptions, beliefs and preferences evolve, guided by the consistency inherent in the underlying theory. Initially, the models used are very simple. But, gradually as intuitions emerge, the models are refined. A cyclic process is followed in which models are built, the output reflected upon and examined for sensitivity, intuitions emerge leading to revision of the models, and so on. This process is stopped when no further intuitions emerge. Larry Phillips terms the final models **requisite**, i.e. sufficiently detailed to bring the decision makers enough understanding to make the decision before them.

The ultimate models are not a description of reality; in a sense, they are reality for the decision makers. Moreover, the modelling process is a dynamic one, in which the perceptions, beliefs and preferences which are modelled change because they are modelled.

Requisite Decision Modelling: A Case Study

LAWRENCE D. PHILLIPS

Decision Analysis Unit, Brunel University, Middlesex

This case study in decision analysis concerns a company that had to decide between continuing to manufacture an old product that might in the near future be banned by the government or introducing an improved but conventional product that would beat the ban but might lose market share to competing products using microchip technology. A decision tree with three attributes describing the consequences over a ten-year horizon modelled the problem. Implementation on a micro computer facilitated extensive sensitivity analyses, the final round of which was conducted by the Board of Directors. More and more pessimistic assumptions were made until the decision switched from the new to the old product: at that point no Director believed all the assumptions. Thus, agreement was reached about the decision even though the Directors disagreed about the uncertainties. The case illustrates 'requisite' rather than optimal decision modelling and shows the essential roles of problem structure and sensitivity analysis.

INTRODUCTION

This case study illustrates 'requisite' rather than optimal decision modelling, shows the essential roles of structuring and sensitivity analysis and indicates how a decision model can be used in a group setting. Although the client had to make a decision in the face of considerable uncertainty, probability assessment turned out to be a relatively unimportant issue.

These features show that this case study departs from the conventional wisdom about decision analysis embodied particularly in textbooks on the subject. In laying the foundations of modern decision analysis, Raiffa and Schlaifer[1] drew on and extended statistical decision theory in a presentation that was heavily mathematical and oriented to optimal modelling. That influence, though much attenuated, can still be seen in recent treatments such as Keeney and Raiffa[2] and is one reason why decision analysis is sometimes criticised as being too mathematical or too inflexible to accommodate real decision problems. Those attempting to apply decision analysis soon discover the difficulty in constructing an optimal model of the decision process, for the very act of modelling often changes the client's understanding of the problem. Thus, real decision making is seen to be labile, a feature that defeats any attempt to build an optimal model of a reasonably stable process.

Attempts to use decision analysis often founder in the early stages of structuring the problem when it is discovered that every person who has a stake in the decision has a different view of the problem and a different opinion about which are the most important features. Texts on decision analysis usually say little about how a single problem structure can emerge under these circumstances, and they often fail to indicate the importance of sensitivity analysis as a method for dealing with differences of opinion. Decision analysis is even portrayed as being applicable only when there is a single decision maker, since the prescriptions of decision theory are applicable only to the preferences of a sincle individual who is striving to be coherent.

Finally, objections sometimes focus on the difficulty in obtaining agreement about the neccssarily subjective judgements of probabilities for future events. For example, Harrison[3] found that many companies do not like to deal explicitly with probabilities, for that only increases the scope for argument and disagreement when recommendations about decisions are referred up the executive hierarchy.

This case study shows that none of these objections is an insurmountable obstacle to implementing successful decision analysis. Instead, most of the obstacles disappear if decision theory is seen as providing a framework for the iterative development of a coherent representation of the problem. In this sense, decision theory is *conditionally prescriptive* in that it polices coherence only within the 'small world' of the problem at hand rather than serving as a normative model for decision making in general.

The goal is to develop a 'requisite' representation, requisite in the sense that everything required to solve the problem is either included in the model or can be simulated in it. To develop a requisite model, it is necessary to involve all those who are in some way responsible for aspects of the decision in the development of the requisite model. The process of building the model is iterative and consultative, and when no new intuitions emerge about the problem, the model is considered to be 'requisite'.

In requisite modelling, it is expected that people will change their view of the problem during the development of the model; that is why the process has to be iterative. Sensitivity analysis plays a key role in facilitating structural change, in helping to resolve disagreements about the implications of differing assumptions on the decision itself and in showing the extent of disagreement about assessments and judgements that can be tolerated for a given decision.

In short, requisite decision modelling treats problem solving as a dynamic process in which all relevant actors become clearer about the problem and develop a deeper understanding of it over time. The model provides a framework for thinking about the problem, thus facilitating analysis. These features of requisite modelling are illustrated with the following case study. The name of the company and its product have been changed to preserve anonymity, and the original monetary amounts have all been multiplied by a constant. All other features have been preserved.

A CASE STUDY

The problem

Maritime Engines and Motors (MEM) is a medium-sized British firm manufacturing outboard engines and motors for pleasure craft. Sales of one particular model are responsible for over half the annual revenues of the Company, and this product has maintained a remarkably steady market share of 70% for over 20 years. It is known for its reliability and sound engineering: it will start even after being totally submerged under water for many hours.

However, this model has recently been banned for use in the inland and coastal waterways of the United States because it does not meet federal emission standards. Although the loss of this market is of little concern since most export sales are to former British colonies in Southeast Asia, the Managing Director of MEM is concerned that it could be banned in Britain. There is no information that the Government has any intention of tightening emission standards, but the possibility cannot be ruled out.

To anticipate the ban, the Managing Director asked for the carburation system of the motor to be redesigned using conventional but up-to-date technology. Tests have confirmed that this product would meet the emission standards of the United States and so would certainly satisfy any new British requirement. Changing to this redesigned product would beat a ban, but it is risky in two respects. First, the new motor looks different from the old one, and so might deter potential purchasers looking for the old, reliable product and who might be unwilling to try a new product untested by time. Secondly, reports have been received that a German firm is about to introduce a new motor of about the same horsepower that uses microchip technology in its carburation system. It is generally recognised in the industry that eventually all engines and motors of this type will use chips, so it would be foolish to introduce a conventional product if the microchip product is to gain public acceptance sooner rather than later. However, the Company sees a shift to the new product as a good opportunity to introduce robotics to the firm. Unions are generally more agreeable to this when a new product is introduced than when it represents only a change to the method of assembling an old product.

The Company was unable to move directly to a chip-based product because it did not have the technology (though a small group was working to acquire it) and the capital investment required was too high (around 30 million pounds sterling). It could just afford the investment in robotics, about 8 million pounds sterling.

In summary, the old engine might be banned, so MEM is considering introducing a new engine that would beat the ban but which is based on conventional technology. This would give the Company and opportunity to introduce robotics to its assembly line. However, the old product is so well-established that any change to its design might cause some loss of market share, and in any event the new product might soon be overtaken by engines using microchip technology.

Procedure

After attending a Brunel Management Programme course in which decision analysis was introduced, the Managing Director of MEM contacted the Decision Analysis Unit at Brunel University to see if decision technology could be applied to this problem. He was particularly concerned to speed up the decision making process in his Company. The last time MEM had considered introducing a new product, a report had been written recommending approval by the Board of Directors. The Board took exception to certain assumptions made in the report and asked for it to be done over. The revised report was submitted to the Board, who took exception to other assumptions, and so this process continued for eleven revisions over twelve months, at the end of which no decision was taken.

At the first meeting, held in early July and attended by the author, the Managing Director, the Business Planning Manager, the Finance Manager and the Production Manager, an outline of the problem was presented. Major uncertainties were identified: these included when and if the current product might be banned, the acceptability of the new product, when microchip-based engines would be dominant in the market, production costs of the new product and its sales price. Company objectives were discussed: these included profitability, company image (failure to keep up with chip technology might affect sales of other products), updating of manufacturing technology and survival. A time horizon of 10 years from October 1979 was established as a reasonable period for evaluating the possible consequences of the decisions, and it was agreed to ignore inflation since it would have about the same influence whatever decision was taken. Financial modelling would use a DCF rate of 10%, current Company policy.

At the end of the discussion, terms of reference were agreed, and these formed the basis for a proposal subsequently submitted to the Company. This was accepted, and the first working meeting was held on 30th July 1979. During August and September a decision tree was developed and refined through the process of showing successive versions of it to a variety of people, who were usually seen individually. They included the Sales, Finance and Production Managers, and some of their staff, and the Managing Director. A luncheon meeting with a key member of the Board of Directors alerted him to the type of analysis being carried out.

By mid-September the Business Planning Manager had completed the financial model. This was married to the decision tree, and the resulting decision model was implemented on an IBM 5110 computer using generic software for building and analysing decision trees. Sensitivity analyses led to further refinements of the model, and by the end of October the modelling exercise was essentially complete. The financial and decision models were explained at a meeting of the Board of Directors held in late November. As had been expected, disagreements were voiced about some of the assumptions used in the model. However, since the computer, which is portable, had been brought to the meeting, it was possible to test the sensitivity of the decision to these different assumptions. This was done, and two hours from the start of the meeting a decision was agreed. It was implemented in January 1980.

The decision model

Figures 1, 2 and 3 show the final decision tree in three segments. Note that the segment shown in Figure 2 is to be attached following path A in Figure 1 and again at path B. The segment in Figure 3 attaches at C, D and E. Thus, the complete tree contains 70 possible paths.

The tree starts with a decision node representing the choice between going for the new, redesigned product or staying with the old one. If the old product is banned, it was assumed that a two-year grace period would be allowed before the ban would be effective for an immediate ban would open the doors to foreign products.

Following the 'new product' decision is an event node that represents the uncertainty associated with all possible sources of trouble in introducing the new product: difficulties with the robotics, inadequate marketing, troubles with the

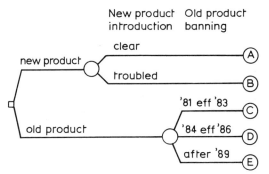

Figure 1 The beginning of the decision tree.

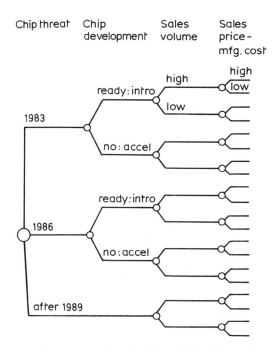

Figure 2 Section of the tree attaching after A and B in Figure 1.

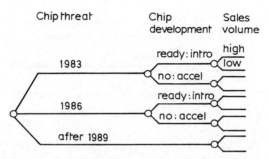

Figure 3 Section of the tree attaching after C, D and E in Figure 1.

engine's design, 'wait-and-see' attitude by the public, etc. It is important to note here that this event node was introduced by the Managing Director himself on the last iteration of the decision tree. He recalled that previous attempts to introduce new products had not always gone smoothly in the Company; this had not been mentioned by anyone else at any previous meeting. As will be seen, the decision is sensitive to uncertainty about this event.

The structure following A or B is shown in Figure 2. The chip threat may develop in 1983, 1986 or after 1989. One of these outcomes is deemed to have occurred if the sales force report more than a 10% penetration of the market by chip-based products. If the threat develops within the 10-year horizon, then either the in-house group working on the technology will be ready with a new engine or it will not, in which case extra funds will be made available to accelerate development. It is assumed, then, that introduction of a chip-based product will be delayed for one or two years. Next, sales volume is assumed to be either high or low; each branch is characterised by a vector of sales volumes (over 10 years) that depends on whether introduction is clear or troubled. The final uncertainty concerns the difference between the sales price and manufacturing cost. The difference was assumed to be either high or low, with amounts between simulated by changing the balance of probabilities between the two branches.

The structure following paths C, D and E, shown in Figure 3, is the same as in Figure 2 except that the final event node in the latter is omitted because both the sales price and the manufacturing cost of the old product are known. However, not shown on Figure 3 are several strategic decisions that would be made if the old product were to be banned. If the ban is announced in 1981, effective in 1983, the Managing Director decided that he would shift to the new product (not the chip-based product) but produce it using conventional methods. This would save the cost of robotics which would be too risky to introduce in the light of the chip threat. If the ban was not announced by 1984, and the chip threat had developed by then, he would immediately commit the Company to shifting to the chip-based product. But if the threat had not developed by the time the ban was announced, then he would recommend manufacturing the new, conventional product on old equipment. Development of this strategic plan occurred when the parts of the tree were put together and the need for a plan became obvious. However, the contingent actions were the subject of considerable debate, which was settled only when the problem was put to the Managing Director.

The decision criteria

Each path through the decision tree contains all the assumptions needed to specify the consequences over a 10-year period. These were evaluated using three criteria: (1) net present value associated with only the old or new product, (2) value associated with the chip-based product, and (3) value of new manufacturing technology.

The first criterion is strictly financial. It was computed by entering all the assumptions associated with a given path into the financial model, a discounted cash flow model that included all capital outlays as well as revenues from the old or new product. By doing this for each path, seventy net present value (NPV) figures were generated. These ranged from about 110 million pounds sterling for the best scenario to about 30 million for the worst.

Contributions associated with sales of the chip-based product, on those paths where this happens, could not be included in the financial model because the capital and manufacturing costs were unknown; not even a prototype had yet been built. Thus, consequences were evaluated under this criterion using judgemental modelling. First, the seventy paths through the tree were found to reduce to only ten distinctly different scenarios with respect to contributions from the chip-based product. Second, the cash flows anticipated for each of these ten were described in qualitative terms. Third, these descriptions were rank-ordered by the Business Planning Manager by considering the qualitative net present value each represented. Fourth, the scenarios were rated on a scale from 0 to 100, with the highest-ranking scenario assigned 100 and the lowest-ranking zero. Finally, this value-scale was mapped to a scale of pounds sterling. Zero on the rating scale corresponded to the scenario in which the Company would not introduce the chip-based product, so it was assigned zero pounds (ordinary development costs were considered as sunk costs and so were not included under any criterion). Even under the top-rated scenario, sales of the chip-based product would not begin until 1986, so the total NPV would be less than half that of the best scenario under the NPV criterion. After some discussion, a figure of 40 million pounds sterling was agreed.

The third criterion was intended to capture the value to the Company of introducing robotics manufacturing technology. All 40 consequences following the decision to stay with the old product were evaluated at zero since robotics would not be introduced. The 20 possible consequences following the clear introduction of the new product were assigned values of 100, while the 20 following troubled introduction were given values of 80 since the robotics might be one source of trouble. This 0–100 scale was then mapped to pounds sterling by considering possible future savings on other product lines as robotics were introduced. It was agreed that this could amount to no more than 8 million pounds sterling; all ratings of 100 were replaced with this figure, while ratings of 80 were assigned 6.4 million.

Finally, because the mappings on the latter two criteria were accomplished by multiplicative factors about which there was some uncertainty, provision was made in the model for changing these factors. This was done using weighting factors on the figures given above. For the base case model, these factors were set to one for the NPV criterion and 0·5 for each of the other two criteria, equivalent to multiplicative factors of 20 million and 4 million per 100 points, respectively, on these criteria.

Probability assessments

Company personnel who had information about the events depicted on the tree were consulted to obtain probability assessments. No special elicitation techniques were used. Discussion about each event culminated in agreement on probabilities to be used in the base case. These will be reported where needed in the following section.

Sensitivity analyses

Many sensitivity analyses were conducted prior to the Board meeting and nothing that had not already been tried was suggested by any director. However, the analyses carried out in that meeting will be reported here because they capture the flavour of the power of this approach.

The meeting opened with an explanation by the Managing Director of the decision tree with a justification for the probability assessments. Then the Business Planning Manager discussed the financial model that had generated the NPV figures, and he outlined the approaches used in evaluating the consequences under the other two criteria. The author then explained the simple mathematics used to combine the criterion values and to roll back the tree (weighted averaging in both cases). The results of the base case model were then shown using the computer: these are indicated within the circles in Figure 4. The new product decision results in an expected value of 81·6 million, the old product in 77·2 million. The difference is not great, so the expected values at A and B were computed; they are 88·8 and 70·4. The old product expected value of 77·2 falls between these two, so the probability of a clear introduction is crucial to the decision. Nobody on the Board thought it should be below 0·5, and nearly everyone agreed that 0·6 was realistic, so it was left at the base-case value.

Attention next turned to the ban. Board members agreed that the probabilities assigned were too optimistic, that more weight should have been given to the upper two branches. However, before these probabilities were changed, the expected values at C, D and E were computed. They are 75·2, 74·0 and 78·0, respectively – not much difference. It was noted that whatever the probability

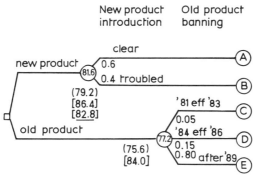

Figure 4 Expected values for various sensitivity analyses.

assignments, the expected value for the old product could not be less than 74·0 nor more than 78·0, both values less than the expected value of 81·6 associated with the new product. Thus, the decision is insensitive to the entire range of possible probability assignments for the ban event. The reason for this insensitivity is that the Managing Director had been forced to think strategically about the consequences of a ban in the light of the microchip threat, and he had devised plans that would cope effectively with the ban. He later stated that he would probably not have thought so carefully about these future events if the decision tree had not been built. Thus, to the surprise of the Board members, the uncertainty that had motivated the analysis to start with turned out to be unimportant.

Since no changes in the ban probabilities were needed, attention next focussed on uncertainty about the microchip threat. Again, Board members felt that the base case was too optimistic, that the probability assignments of 0·2, 0·3 and 0·5 to 1983, 1986 and after 1989 events gave too little weight to the earlier dates. A survey of the Board revealed that the most pessimistic individual would assign values of 0·5, 0·3 and 0·2, respectively. These values were entered into the computer, which changed the assignments at each of the five places where this node occurs, and the expected values associated with the decisions were computed. While the computations were being carried out, Board members were invited to judge the effects of this change in probabilities. Most believed that the new product would become less attractive, and that the old product might be preferred. The new expected values are 79·2 and 75·6 (shown in parentheses in Figure 4). The new product decision had become less attractive, but so had the other one. On reflection, this is reasonable, since the microchip poses a threat to the old product as well as to the new one.

Next, the Board felt that too little weight (0·5 in the base model) had been given to the second criterion, value associated with the chip-based product. Increasing the weight to 1·0 gave expected values for the decisions of 86·4 and 84·0 (in square brackets on Figure 4); the new product decision was still better, but the gap had narrowed to 2·4 million.

At this point, the Board were wondering what assumption had to be made to damage the case for the new product. One member noted that probabilities of 0·8 and 0·2 had been assigned to the high and low branches of the event representing the difference between sales price and manufacturing cost, given clear introduction of the product (the probabilities are reversed if introduction is troubled). These were then reversed for all branches following clear introduction, and the expected values of the decision options recomputed. Only the expected value for the new product changes; it was 82·8 (bracketed and underlined in Figure 4). At last the decision had switched; the expected value of 84·0 associated with the lower branch was higher.

However, Board members were reminded that all changes had been cumulative, so that the decision to stay with the old product should be adopted only if everyone agreed with the more pessimistic probabilities associated with the ban and revenue events and with the increased weight on the second criterion. Nobody did. Each person would relax at least one set of assignments back to the base case, and that would lead to the new product decision being preferred. Thus, although they did not agree about the assumptions, the Board did agree to go ahead with the new product.

DISCUSSION

In the course of the analysis, the Managing Director noted the difference of 11·6 million between the expected value associated with the old product in the base case (77·2) and the expected value following a clear introduction (88·8). He realised he would be justified in spending up to 11·6 million to ensure a clear introduction and that though this goal was impossible he could still do a great deal to increase the probability above 0·6. Eventually, he called in an outside design consultant to redesign the new product so it would look as good as the old one, and he spent several thousands of pounds to ensure that the marketing of the new product was done as well as possible. Thus, he used the analysis to justify spending these extra amounts of money, which were well below the maximum of 11·6 million; the analysis helped him to think clearly about the problem and called his attention to the importance of a clear introduction, an event that had not even been mentioned at the first meeting.

The analysis also helped the Board to clarify their thinking about which events were important; they had all been worried about the possible effects of a ban, but this was shown to have no influence on the decision, largely because the Managing Director had devised an effective strategy to cope with the ban. Furthermore, the Board had not realised that the microchip threat affects both decisions. The more usual approach of the Board would have been to discuss the pros and cons of a particular recommendation, in which case they would certainly have seen the microchip threat as a risk only to the new product, so using this as an excuse not to go ahead with the new engine.

Note, too, that if the Managing Director had not been a part of the iterative development of the decision tree, a crucial event would have been left out. This could have endangered acceptance of the model if the event had come to light in the Board meeting. The Managing Director is more likely to take a strategic view of the Company than any of his unit managers, so it is important that he be involved in the development of the decision model, even if he has time only to pass comment on the state of the model. Thus, involvement on the part of the individual who is accountable for the decision is an important ingredient of a successful application of decision analysis.

Perhaps the most important point to be drawn from this case study is that decision analysis was used not so much as an optimal model that prescribes the 'correct' course of action, but rather as a framework for thinking about the problem, for exploring the consequences of different assumptions. More particularly, for determining the set of assumptions needed to switch the decision. Exploration of the reasonableness or otherwise of those assumptions is then done outside the model. Thus, the model becomes an aid to decision makers who then decide for themselves which course of action to adopt.

In a sense, no model of a decision problem can ever be optimal. Where problem solving is concerned, there is no reality to be modelled; the model is the reality. Different people have differing views of a problem, and as they discuss it, they change and modify their internal representations of it.[4] In carrying out a decision analysis, one attempts to construct, to generate an explicit representation of the problem. It would be foolish to say that any particular representation is optimal, for there can be no criterion against which to make that judgement. That is why the term 'requisite' model has been introduced here. A model is considered requisite when no new intuitions arise about the problem, at least no intuitions that would affect the decision. Requisite modelling exploits the insensitivity of the decision to many of the factors that may at first be judged to have an influence on the decision. Thus, a requisite model is not necessarily a 'satisficing' model,[5] nor is it an optimal model that is degraded because the cost of thinking is too great. Instead, it is a representation of the problem deemed sufficiently adequate by the decision makers to provide them with a useful guide to thinking about the problem.

ACKNOWLEDGEMENTS

Appreciation is extended to T. K. Wisniewski and S. Wooler, who, with the author, formed the team of decision analysts. The splendid cooperation of personnel at the Company contributed in great measure to the success of the project. Thanks, too, to Decisions and Designs, Inc., who supplied the computer program.

REFERENCES

[1] H. Raiffa and R. Schlaifer (1961) *Applied Statistical Decision Theory*. Graduate School of Business Administration, Harvard University.

[2] R. Keeney and H. Raiffa (1976) *Decisions With Multiple Objectives*. Wiley, New York.
[3] F. L. Harrison (1977) Decision making in conditions of extreme uncertainty. *J. Mgmt Stud.* **14**, 169–178.
[4] R. M. Hogarth (1981) Beyond discrete biases: functional and dysfunctional aspects of judgmental heuristics. *Psych. Bull.* **90**, 192–217.
[5] H. A. Simon (1955) A behavioral model of rational choice. *Q. J. Econ.* **69**, 99–118.

3.7 SELECTION FROM A SHORTLIST

So far as we have described it, decision analysis has focused on choices in which there is significant uncertainty. It has been assumed that at the moment of choice some exogenous factors, which comprise the true state, are unknown. Moreover, the consequence of the decision depends on their true values. There are, of course, decision problems in which uncertainty is not important, essentially because all the determinants of the consequence are known once the decision is made. In a sense, such **decision problems under certainty** are easy. One simply chooses the action that leads to the most preferred consequence. (For this reason, often actions and consequences are not distinguished in these problems.) This view, however, is too simplistic.

Usually decision makers need the support of decision analysis to sort their thoughts and feelings out in two major respects. First, they need to weigh up the uncertainties before them. Second, they need to consider their preferences about often complex consequences. When they face a decision problem under certainty, they naturally do not need help in the first respect; but their need in the second respect remains. Structuring their descriptions of the consequences according to a hierarchy of attributes and evaluating their preferences through a multi-attribute utility function can be enormously helpful.

Valerie Belton's paper, which won her a Mike Simpson Award in 1984 and was published in 1985, describes a case study in which this was so. Before pointing to particular points in her paper, it will pay us to pause and consider decision problems under certainty more generally.

Multi-attribute utility functions encode two distinct aspects of preference: (i) attitude to risk, and (ii) trade-offs between levels on different attribute scales. The first of these aspects is clearly irrelevant to decision-making under certainty. Thus the process of assessing a multi-attribute utility function would require more effort from the decision makers than was strictly necessary. Furthermore, that effort, involving as it does hypothetical bets, might well confuse decision makers facing up to a choice with no uncertainty.

To make this point more specific, consider the form of utility function that arises in the case of preferential independence of two attributes (section 2.4, p. 58):

$$u(x, y) = r(v(x) + w(y))$$

Here the function $r(\cdot)$ encodes attitude to risk, while the functions $v(\cdot)$ and $w(\cdot)$ together encode the trade-offs between the two attributes. In circumstances of certainty, the function $r(\cdot)$ is irrelevant and one may rank the actions according to:

$$V(x, y) = v(x) + w(y)$$

(Remember that $r(\cdot)$ is strictly increasing and so $u(\cdot,\cdot)$ produces the same rank order as $V(\cdot,\cdot)$.) Thus in such a case one would assess $v(\cdot)$ and $w(\cdot)$, but not $r(\cdot)$.

Functions which represent preferences for multi-attribute consequences without encoding an attitude to risk are known as **multi-attribute value functions**. The underlying theory is discussed in detail in French (1986) and Keeney and Raiffa (1976).

There is another respect in which the case study in Belton (1985) differs from the theory that we have presented. We have suggested that consequences are first described by a vector of attribute levels and then evaluated in terms of preference through a multi-attribute utility or value function. In practice, it is often possible to combine these two steps.

Continuing with the two-attribute, preferentially independent example above,

$$V(x, y) = v(x) + w(y)$$

the functions $v(\cdot)$ and $w(\cdot)$ may be thought of as bringing preferences for the two attributes onto the same scale (in a very loose sense!; see French, 1986). When two consequences (x_1, y_1) and (x_2, y_2) are indifferent:

$$V(x_1, y_1) = V(x_2, y_2)$$

So:

$$v(x_1) + w(y_1) = v(x_2) + w(y_2)$$

and, hence:

$$v(x_1) - v(x_2) = w(y_2) - w(y_1)$$

That is, the difference in value as measured by the $v(\cdot)$ scale brought by the difference in levels on the X-attribute equals the difference in value as measured by the $w(\cdot)$ scale brought by the difference in levels on the Y-attribute. Thus the two scales measure preference in the same units in the sense that one unit on the $v(\cdot)$ scale trades off with one unit on the $w(\cdot)$ scale.

Now, providing that the decision makers can think in terms of equal differences in preference on the attribute scales, it may be possible to assess $v(x_1)$ directly, rather than measure the level x_1, assess the whole function

$v(\cdot)$, and then combine them to give $v(x_1)$. Roughly, one asks the decision makers to position a consequence on a scale of, say, 0 to 100 points with respect to their preferences on the X-attribute. Belton gives brief details.

Two final points before I leave you to read the paper. First, we have used the word 'attribute' consistently. 'Criterion' and 'objective' are also used. The latter usually means an attribute for which the greater the level, the greater the preference: one tries to maximize an objective. Second, Belton questions whether her analysis is requisite in the sense of Phillips. Despite her conclusion that it is not, I cannot but help feel that she and Phillips use different words to describe the same ideas.

The Use of a Simple Multiple-Criteria Model to Assist in Selection from a Shortlist*

VALERIE BELTON
University of Kent at Canterbury

The use of a simple multiple-criteria model to assist in decision making is described. The model, a hierarchical additive weighted value-function, was used as a part of a decision-making process to select, from a shortlist of three, the company with which to place a contract for the development of a computerized financial management system. The multiple-criteria model and its use are described in detail. To conclude, there is a discussion on the contribution of the model to the decision-making process as perceived by the decision-making group.

Key words: decision, multi-objective

INTRODUCTION

This paper describes the use of a simple multiple-criteria model as a decision aid in a large service company, which will be called Financial Information Services Ltd (F.I.S.L.) engaged in the process of choosing a company with which to place a contract for the development of a computer system. The computer system, a financial management aid, was to be provided to clients as a chargeable service.

F.I.S.L. did not have expertise in-house to develop a proven and reliable system of the type it wished to market sufficiently quickly to respond to pressure from competitors. By involving an outside organization, it hoped to be able to overcome these difficulties whilst meeting the objective of educating its own staff in this type of system. A further possible objective was to bring the system in-house at some time in the future. The system is a large-scale one involving the extraction and

*The original version of this paper was one of the 1984 winners of the Mike Simpson Awards and was presented at the *O.R. Society National Conference*, University of Lancaster, September 1984. The Awards were instituted in memory of Professor M. G. Simpson of the University of Lancaster, President of the O.R. Society 1978–1979.

collation of data from worldwide sources and the formatting of the data into reports for use by clients anywhere in the world.

THE DECISION-MAKING GROUP

The project is a large-scale one, for which development costs were estimated to run into several millions of pounds over the initial 2 or 3 years. Thus, to ensure that due consideration was given to the choice of organization with whom the contract for the development of the system was placed, a working group of specialists was formed. The working group was 12 people strong and comprised mainly systems analysis staff and marketing staff of F.I.S.L. Other specialist advice, for example on computer audit, legal matters or costing issues, was sought as necessary. The working group was responsible for making a recommendation to an executive committee of senior managers of the company, with whom the responsibility for the final decision rested.

THE DECISION-MAKING PROCESS

An initial longlist of potential companies to fulfil the contract was reduced to a shortlist of three for detailed consideration. A proposal was submitted by these three companies in response to a request which posed in excess of 2000 questions relating to nine major areas of concern, which will be described later. It was the task of the working group to evaluate these detailed proposals and to form a recommendation. The evaluation was a thorough one. To ensure a fair representation of each of the companies who had submitted proposals, a number of procedures were adopted, one of which was the simple multiple-criteria analysis which will be described in detail later. The process of evaluation followed two lines: first a general evaluation, and secondly, to complement this, positive and negative reviews. The general evaluation made use of the multiple-criteria model to identify key issues which differentiated between the companies and to focus attention on these issues. A hierarchical, weighted additive model was used to assess each of the companies in the light of the objectives of the project as described by the nine major areas of concern. A comprehensive sensitivity analysis was carried out, concentrating on the strengths and weaknesses of the companies as identified in the key issues and positive and negative reviews. To ensure a fair positive review for each company, an internal advocate was appointed to each. This was a member of the working group whose responsibility it was to work closely with the appointed company and to support their case in meetings of the working group. This procedure had the additional effect of encouraging adversarial activity in the decision-making group, which is generally thought to be beneficial to the decision-making process. L. D. Phillips[1] states "Experience with decision conferences suggests that the adversarial process helps individuals to broaden their individual perspectives on the problem, to change their views,..., in short, to create a model that fairly represents all perspectives". Such benefit may also result from the negative review procedure, the major function of which was to highlight potential major problems and to allow the companies the opportunity to indicate how they might deal with such. At this stage, all three companies appeared to be sufficiently flexible to allow for such difficulties to be overcome.

With the benefit of all the information contained in the proposals, supplemented by a number of further review meetings with each of the companies, the working group retired for an intensive 3-day working meeting to carry out a final analysis and to formulate its recommendation to the executive committee. It was during this meeting that greatest use was made of the multiple-criteria model, particularly for sensitivity analysis.

The final outcome, after considerable debate, was a unanimous decision on the company to receive the contract.

ANALYSIS USING THE MULTIPLE-CRITERIA MODEL

Statement of objectives

As was mentioned briefly in the Introduction, the objectives of F.I.S.L. in employing an outside agency to develop the system for them were as follows:

'Education'. F.I.S.L. had little in-house knowledge or experience of the type of system being developed and thus required assistance initially.

'A proven product'. It was felt desirable to have a reliable system from the outset to offer to major clients. F.I.S.L. wished to avoid the need for experimentation in introducing the system.

'Timing'. In order to respond quickly to competitive pressure, it was necessary to have a working system available in the near future.

'Development'. While recognizing the need for involvement of an 'expert' company initially, one of the long-term aims of F.I.S.L. was to bring the system in-house, and the chosen collaborative company should be one which could progress towards that goal.

Criteria for evaluation

Whilst the above represented the broad objectives of F.I.S.L., it did not immediately present a workable framework for detailed evaluation of the companies. Had time been available, it would have been possible to develop such a framework around these objectives; however, the multiple-criteria model was not a fundamental building block of the analysis but was adopted in response to the need to analyse the information contained in the proposals. Thus the detailed evaluation was based on the proposals which had provided information on nine major areas of concern. These nine areas are listed below:

(A) Corporate status and structure. The size and standing of the project was such that it could only be supported by a company of substantial financial strength.

(B) Geographic coverage. When in use, the system would require access to data held by F.I.S.L. at numerous centres worldwide. Thus it would facilitate development if the chosen company had a developed network coincident with that of F.I.S.L. and their proposed future development.

(C) Systems issues. This section is divided into many sub-criteria, all relating to questions about the company's computer systems, the hardware, software or communications aspects: for example, the nature of the communications network, its capacity, availability and security, the design of the system software.

(D) Marketing issues/Service features. As the system was to be developed for F.I.S.L. to sell to its own clients, it was important to have a satisfactory product to offer. Thus it was necessary to consider questions such as the user friendliness of the system, the company's strategies as regards terminals to access the system, the extent to which each company would allow customization of report writing to suit F.I.S.L.'s perceptions of their clients' needs.

(E) Ease of implementation. In line with the overall objective of having a working system available quickly, it was important that implementation was well planned and seen to be efficient.

(F) Enhancements. F.I.S.L. was concerned that if future enhancements to the system should be required, they could be achieved as easily as possible in view of likely competing priorities with other customers of the contracted company.

(G) Issues relating to costing and pricing of the service. These considerations were important as different pricing strategies; for example, one based on the use of C.P.U. time vs one based on the number of items of data reported could result in very different costs to the client. However, evaluation of these issues was fraught with difficulties because of the uncertain nature of forecasts of usage of a new system. In the light of this, a flexible attitude towards costing was considered to be desirable.

(H) Contractual issues. There were important issues relating to both the style and the negotiation of the contract. There were potential difficulties here because of the international nature of the system and because one of the companies under consideration was not U.K. based.

(I) Strategic issues. Again, a broad range of issues was considered. Of particular importance was the likely case of transfer of the system from the contracted company to in-house.

These criteria formed the second level of the multiple-criteria hierarchy used for evaluation. The first level of the hierarchy was a composite overall objective which was never stated explicitly but was a function of the four fundamental objectives stated at the beginning of this section. Some of these second-level criteria were broken down further into a third level of criteria to complete the multi-attribute hierarchy. Where it was appropriate, the identification of the more detailed criteria

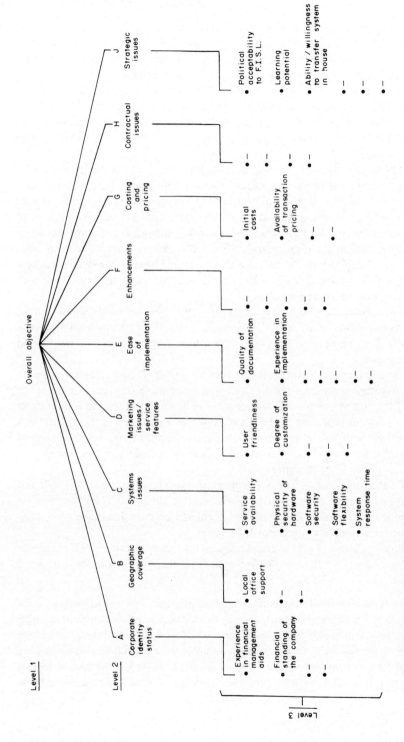

Figure 1 The hierarchy of objectives.

was done by those members of the working group with expert knowledge in a particular area; for example, systems issues were considered by the computer systems people. The complete hierarchy is illustrated in Figure 1, although it has not been possible to state all criteria explicitly because of the need to preserve confidentiality.

Evaluation

The detailed evaluation of each of the companies with respect to the criteria identified above and detailed in Figure 1 was carried out by those with expert knowledge in the area, where appropriate, and drawing on the knowledge of the working group as a whole for the other criteria. An individual was appointed to ensure the best use of expert knowledge in the evaluation with respect to each level-3 criterion. Before commencing this stage of the evaluation, this individual, the 'leader', for each section was asked to review the criteria once again, adding any new issues that had arisen and deleting any which had become redundant. The next step was to consider each criterion carefully and to identify any which were really constraints – i.e. criteria on which a minimum standard need be attained and beyond which no further differentiation is relevant to the contribution to the overall objective. Such criteria were not included further in the multiple-criteria model.

No formal checks were made of the independence assumptions necessary for an additive model of this type to be valid, but we were aware of these in the specification of the model and throughout the analysis. It is generally accepted that the additive model is a robust one,[2,3] and it was felt that greater insight would be gained in the time available from the use of a simple model than to attempt to construct a more complex representation of the decision.

The evaluation of the three companies with respect to the level-3 criteria was on a 0–100 scale. The anchor points represented the company which did worst and best on that criterion. Thus, having ordered the companies in terms of their performance on a particular criterion, it remained only to specify the position of the middle-ranked alternative relative to the best and worst. No attempt was made to define physical measurement scales for any of the criteria. Although such may have been obvious for some criteria, for the majority it would have been a difficult and time-consuming task, the value of which was not apparent at this stage. Indeed, at a later stage, sensitivity analysis suggested that such detailed information was not called for. Thus the working group was required to assign a value representing any non-linearities of values scale which may exist for those criteria which related easily to a physical measurement scale, and for the other criteria, a direct subjective rating. If any criteria emerged at this stage on which the companies were not differentiated, that is, they were considered to perform equally well with respect to those criteria, a note was made of this and an appropriate zero weighting was applied at the next stage of the analysis. In practice many criteria fell into this category, and the size of the 'effective' multiple-criteria model was consequently considerably reduced.

Having scored the companies with respect to all the criteria at level-3 of the hierarchy, the next stage was to weight those criteria to reflect their relative contribution to the criteria at level-2. This was done using a variation of the 'swing weights' procedure, described in detail in Keeney and Raiffa.[4] The object of the weighting procedure is to scale the criteria scores appropriately, that is, to assess the relative worth of 100 points on each criterion. It is, of course, important to relate the criterion scales to the options at this point and to appreciate the interaction between weights and scores. The differential between the lowest and highest-ranked companies on one criterion may be minimal, but may be very great on another criterion, although in each case the differential is represented by 100 points. The weight assigned to a criterion must reflect the significance of this differential as well as the importance of the criterion. Thus, a wide differential on an important criterion signifies a very high weight, whilst a narrow differential on an unimportant criterion should signify a negligible weight, and so on. Much emphasis was placed on the need to understand the meaning of the weights and their interaction with the scores, and care was taken to explain it in detail to each member of the working party. It was suggested that they identify the most important criterion and assign to it a weight of 100 and then determine the weights of the other criteria relative to this. In the final analysis, which was done by computer, the weights were normalized to sum to 1.

The next stage was to weight the criteria at level-2 in line with their contribution to the overall objective. The weights at this level represent the cumulative weight of all level-3 criteria which are sub-criteria of a particular level-2 criterion. These weights can be assessed either by direct comparison of the criteria at level-2 or by selective comparisons of criteria at level-3. It was decided to adopt the former, more direct approach and to supplement this by consistency checks using the implied level-3 weight. Initially the persons responsible for each section of the analysis were asked to give their personal opinion on the weights to be used at level-2. This information was used as the basis for sensitivity analysis before the group came together to discuss this issue. This approach avoided the possibility of long discussions about these weights when the disagreements may have been inconsequential.

Results of the multiple-criteria analysis

A program written by R. Bromage for the HP85 microcomputer was used to analyse the information. The results of the initial run were informative, but the main strength of the program was the ease with which sensitivity analysis could be carried out and the attractive graphical presentation of such analysis. The initial analysis showed that one of the three companies being considered performed very poorly in comparison with the other two. This was demonstrated very clearly by a simple graphical presentation of the information in the multiple-criteria hierarchy aggregated to level-2, the nine criteria representing areas of major concern. This is displayed in Figure 2; each vertical line represents a level-2 criterion, and the

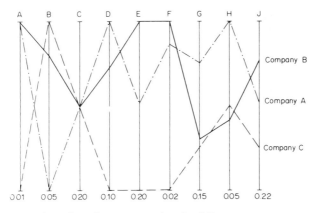

Figure 2 Presentation of results aggregated to level-2.

aggregate score (which can range from 0 to 100) of each company is displayed on each criterion. The three companies will be referred to as Company A, Company B and Company C to preserve anonymity. The figure below each criterion axis is the proportion of weight assigned to that criterion. We see that Company C performs best on only one criterion, geographic converage (B), which was initially assigned only 5% of the total weight. Moreover, on all other criteria but one, criterion H, contractual issues, it was equalled or bettered by both A and B. It must be remembered at this stage that the level-2 criteria are aggregates of level-3 criteria, and that a graph such as this does not necessarily reflect dominance; i.e. a change at level-3 could change the whole picture. Again, sensitivity analysis was used to investigate the possibility of this happening. This confirmed the feeling that Company C was an inferior option in terms of F.I.S.L.'s objectives.

Thus it seemed that the real decision was to be made between Company A and Company B. The overall weighted score emerging from the multiple-criteria analysis indicated a preference for Company B. The scores were 59 and 68 for Company A and B respectively; Company C had scored only 27. However, too much emphasis should not be placed on these numbers. Further inspection of Figure 2, the graph of results aggregated to level-2 of the hierarchy, shows that there are criteria of greater and lesser importance on which each company is ranked more highly. A thorough sensitivity analysis was carried out to identify those changes in inputs to the model which would significantly affect the outcome – i.e. which would reverse the ordering of companies A and B in the overall evaluation. The working group was confident about the evaluation of the companies on the majority of criteria, although there were a few areas for which information was lacking which were considered in detail in the sensitivity analysis. Thus, more effort was concentrated on the effect of changing the weights assigned to criteria at level-2 and at level-3 of the hierarchy. The results of this analysis were presented graphically, as illustrated in Figure 3, for the level-2 criteria E, ease of implementation, and G, costing and pricing issues. Each graph is a plot of the total weighted score of each company against the level-2 weight assigned to the criterion under consideration. The vertical line away from the axis represents the status quo; that is, as seen in Figure 3(b), criterion G currently has a weight of 0·15 and the overall scores of companies A, B and C are 59, 68 and 27, respectively. We see that if we were to increase the weight on this criterion to 30% of the total weight, the ordering of companies A and B would be reversed. By using the HP85 linked to an HP flat-bed plotter, which can be used to draw up viewfoils for an overhead projector, this information could be presented without delay to all members of the working group. This enabled instant discussion of whether or not the change in weight from the status quo to the point at which the decision was reversed was a

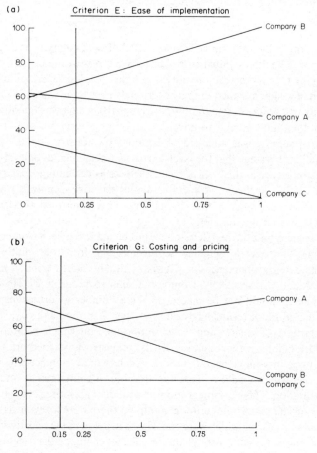

Figure 3 Sensitivity analysis.

reasonable one. In fact it emerged that no such changes in individual weights could be justified; i.e. Company B remained the one with the highest overall score within the range of variation of weights considered to be reasonable.

We were aware that such an analysis is necessarily simplistic and that if more than one weight or score were to be changed simultaneously, the outcome may be more marked. No formal analysis of such possibilities was performed, it being beyond the scope of the computer program, but they were considered in an *ad hoc* manner. That is, by inspection, those multiple changes which could cause Company A to be rated more highly than Company B were identified and considered. None was thought to be reasonable.

Thus, at this stage of the evaluation, it appeared that Company B was emerging as preferred, with Company A a close second and Company C a distant third. This was in accord with the feelings of the working group.

Final evaluation

A final stage in the evaluation was to return to the four broadly defined objectives, education, a proven product, timing and development, and to perform a holistic evaluation of the companies in the light of these objectives. This reconsideration of the fundamental issues confirmed the pre-eminence of Company B. It was felt that Company B was able to satisfy F.I.S.L. on all four objectives. Company A, however, was unable to provide the same degree of education in the development, use and implementation of such systems as Company B, nor was a development path to bring the system in-house so clearly defined. Company C was clearly the weakest candidate in all respects, as measured by F.I.S.L.'s fundamental objectives, although its proposal was a perfectly sound one.

The decision to recommend Company B was not shaken by further recourse to the negative review of each company.

RECOMMENDATION

In the light of the above described analysis, the working group placed a firm recommendation with the executive committee that a contract should be placed with Company B. The recommendation included a suggested budget and a number of specific contractual points.

This recommendation was accepted by the executive committee, and Company B was contracted to develop the system which is now in use by a number of major clients.

DISCUSSION OF THE USE OF THE MULTIPLE-CRITERIA MODEL

The use of such an aid to decision making was new to all members of the working group, although several individuals had had some training in the use of quantitative methods for decision making as part of a management training course or M.B.A. and were receptive to such modelling approaches.

Certain reactions were clear during the course of the evaluation. However, as my research work at the time was concerned with the practical use of such models. I circulated a questionnaire to members of the working group to elicit reactions to the use of the multiple-criteria model. This was done shortly after the evaluation had been completed and the recommendation had been sent to the executive committee. A copy of the questionnaire is attached in the appendix. The questionnaire is in three sections: the first considers the aims of the use of such an approach, the second section contains more specific questions about the model itself and the third a collection of general questions.

The aims of the approach

The statement of these aims is based on a paper by Hinloopen *et al.*,[5] who has made extensive use of multiple-criteria models in regional planning.

All the responses commented on the value of the use of the model in providing a useful framework for the discussion of the problem, the synthesis of much factual information and the incorporation of value judgements about the relative importance of different fators. One comment was made, with which as an observer I agree, that the model was constructed to make use of the information available from the proposals rather than, as would be ideally the case, being structured with the decision foremost in one's mind. However, it was not felt that this had led to the omission of any important considerations, and there was a strong feeling, expressed by all, that its use in this manner had concentrated attention on the information in the proposals which was most relevant to the decision, that is, the differentiating criteria. A personal view is that this had the effect of leading to the apparent disjointedness between the four fundamental objectives and the detailed multiple-criteria evaluation. Whilst the two separate analyses, the detailed evaluation and the holistic evaluation were, I believe, both valuable contributions, it may have been more appropriate and natural if the four fundamental objectives had featured as one of the levels of the multiple-criteria hierarchy.

It was also felt that the use of the model was beneficial in providing a more justifiable basis for the decision, particularly in enabling the working group to justify its recommendations to the executive committee. This does not mean that the model was viewed as merely a scientific justification of a decision which had already been taken. It was clear during the discussions on the evaluation of the companies that it was not an attempt to justify a prior decision but more of a learning process, an exploration of the overwhelming wealth of information which had been gathered and a moulding of this information to a useful decision aid. The appointment of internal advocates to each company had essentially precluded the possibility of a foregone conclusion, although possibly by artificial means.

The nature of the model

The comments in this section related to specific aspects of the model, the scoring and weighting procedures and the presentation of results. It was generally felt that the graphical presentations of results were one of the particular benefits of the process.

Several people were uneasy with the use of a 0–100 scale, with the worst and best options constrained to be 0 and 100 respectively. This is a problem I have encountered when using this scale in other applications of such a model. The difficulties are exaggerated by having only three options for evaluation and thus only one free point. However, the scale is one which is commonly used and these feelings of unease are generally overcome with familiarity.

General comments

A number of suggestions for minor changes were made – in particular, speeding up the program. The HP85 is somewhat of a dinosaur amongst micros, speed being severely restricted by the use of cassettes, but analyses were available quickly if not instantaneously. This comment reflects the fact that faster and more powerful micros can indeed make a valuable contribution to managerial decision making.

CONCLUDING REMARKS

This paper has described the successful use of a simple multiple-criteria model in aiding a decision-making process. The application was perhaps unusual in that the decision process was not developed within that framework from the beginning; however, there is no doubt that the model served the purpose of allowing the decision-making group to explore and learn about the problem and their preferences. It was a straightforward application in the sense that very little conflict arose between the recommendations of the model and the intuitions of the decision-making group. This does not mean that the model was superfluous; it played an important part in enabling the group to understand and justify their intuitions. For some of these involved, this may have been achieved through no more than effective graphical representation of information; for others the synthesis of technical and subjective judgements, and in particular, the extensive sensitivity analysis of the latter, were valued contributions.

The model was never viewed as prescriptive or normative by the decision-making group; neither was it a descriptive model, nor a requisite model, as described by Phillips[1] – "a model whose form and content are sufficient to solve a problem". Rather, it was model which facilitated understanding of the problem, but only as a part of the decision-making process.

APPENDIX

Cambridge University Engineering Department
Control and Management Systems Division
Mill Lane
Cambridge CB2 1RX

Multi-Attribute Value Function Decision Model

Now that you have had time to reflect on the decision-making process, but before it slips from your mind, it would be very helpful to me if you could comment on the use of the multi-attribute value approach, including the use of the computer program. Please be completely honest; positive comments are obviously welcome and encouraging, but criticism is generally of greater value in improving existing approaches and developing new ones. I have not asked many direct questions, as the Yes/No answers they tend to produce are of limited value. Instead, I have suggested a number of areas for comment; you may wish to identify other specific points.

Please give your responses to Andrews, or send them directly to me at the above address. You may respond anonymously if you wish. If you wish to discuss your comments, then I would be pleased to do so; my telephone numbers are Cambridge (0223) 66466 Ext. 347 and 302 and Royston (0763) 41933.

Yours sincerely,
Mrs Valerie Belton
26 August 1982

Section 1: Aims

It has been suggested that multi-attribute approaches to decision making should serve the aims listed below. Could you please comment on the extent to which the decision model and associated decision process helped to achieve those aims? It is your personal view I seek. It would be useful if you could say if you had already achieved each aim before the introduction of the decision model/process and how/if it would have been achieved without the use of the model. Please feel free to disagree with any of the aims listed and to add any other ones you think are relevant.

Aims

1. Surveyable classification of factual information.
2. Better insight into the various judgements (e.g. the assessment of the relative importance of the different criteria).
3. Inclusion of differences in interest and/or political views.
4. Emphasis on the openness of the decision-making process.
5. Meaningful reduction of the available information.
6. Aid to better considered decisions.

7. Provision of a more justifiable basis for decisions.
8. A more structured decision-making process.

Section 2: The Process

You may have felt that difficulties arose at particular stages of the process, or that some aspects were more useful than others. Please make any comments you feel are appropriate. I have suggested a division of the process to jog your memory and to guide you in your responses.

1. The structuring of the decision, i.e. identification of 'sections' and of relevant criteria within sections.
2. Ranking options on the individual criteria.
3. The use of a 0–100 scale.
4. The 'scoring' of the intermediate option.
5. Assigning 'weights' to the criteria within sections.
6. Assigning 'weights' to sections.
7. Display of results – in table form and/or in graphical form.
8. Sensitivity analysis and display of sensitivity analysis.

Section 3: General

1. When, if at all, did you begin understanding the process? From its introduction? At some stage during the evaluation? When the results were presented? Do you still not understand it? (If that is the case, it would be very useful if you could identify why you are unable to understand it; it may be simply because of a lack of adequate explanation. This would facilitate future developments to make it more transparent.)
2. Can you suggest ways in which the process could be improved?
3. Do you think it would have been beneficial to have had an impartial analyst to conduct the process?
4. This approach to decision making has a wide range of potential applications to decisions for which there are multiple and possible conflicting criteria to take account of. Examples range from personal decisions such as buying a car or a house, to many executive/business decisions, in particular those associated with the assignment of priorities and allocation of resources. Would you use a decision-making procedure such as this in any of these instances? I would be interested to hear your views on the general applicability of the approach.
5. Finally, do you have any other comments?

THANK YOU FOR YOUR HELP.

REFERENCES

[1] L. D. Phillips (1984) A theory of requisite models. London School of Economics and Political Science, Decision Analysis Unit.

[2] P. Humphreys and W. McFadden (1979) Experiences with MAUD. Paper presented to the 7th Conference on Subjective Probability Utility and Decision Making, Göteborg.

[3] W. Edwards (1978) Use of multiattribute utility measurement for social decision making. In *Conflicting Objectives in Decisions* (D. E. Bell, R. L. Keeney and H. Raiffa, Eds). Wiley, New York.

[4] R. L. Keeney and H. Raiffa (1976) *Decisions with Multiple Objectives: Preferences and Value Tradeoffs*. Wiley, New York.

[5] E. Hinloopen, P. Nijkamp and P. Reitveld (1982) The regime method: a new multiple criteria technique. In *Essays and Surveys on MCDM* (P. Hansen, Ed.). Springer, Berlin.

4

Thinking and sharing perceptions

4.1 A COLLECTION OF 'VIEWPOINTS'

In section 1.3 I suggested that decision analysis has often been misinterpreted by the operational research profession and, in consequence, several quite inappropriate criticisms have been levelled against it. In particular, it has been assumed that the essential modelling of beliefs and preferences was intended to portray the decision makers' feelings as they existed before the analysis began. The intention that the analysis would provide a framework in which the decision makers could think about, modify and, indeed, evolve their beliefs and preferences was not recognized. Moreover, the central role that decision-analytic models can play in structuring discussion and helping group communication also went unrecognized.

There have been many debates in the general operational research/management science literature in which the role and purpose of decision analysis have been discussed. Adelson and Norman (1969) and the response by Croston and Gregory (1969) summarize one such debate. Below are reprinted a more recent exchange of 'viewpoints' from the pages of the Operational Research Society's journal (Adelson and Tocher, 1977; French and Tocher, 1978; French et al., 1978). It began with a comment made by Tocher on utility theory in his 'Notes for discussion on "Control"' (Tocher, 1976, pp. 233–235).

UTILITY THEORY

The arguments against utility theory are so obvious that they are rarely aired. The concept has such a deceptive attraction for the naive, that it is as well to re-rehearse the fallacies of this pseudo-scientific concept and that of subjective probability at regular intervals.

First, the general thesis: Utility theory hinges on the observation that humans can often rank their preferences. Before we can map this preference onto a number scale, we need an indifference principle. This is done by asking for a preference choice between a single outcome and a compound outcome consisting of a probability weighting of two simple outcomes.

The concept of probability to be used is never clearly defined. If it is a frequency concept then the whole theory fails to recognize that preferences are changed by events. If I am faced with the choice of an apple or an orange I might elect for an apple but by the time I have my 1000th apple, I might pay not to have any more. The frequency concept cannot be applied even hypothetically; attempts to use it generate a circular situation; the preferences would change during the (infinite) experiment to determine the probability.

Thus a subjective concept of probability must be implied and the two theories are tied together. A similar argument is applied in the latter theory. People can rank the strength of their belief in the truth of a proposition and then a measure can be invented if we design an indifference mechanism.

The real appeal of these theories lies in their taking us back to the familiar ground of quantities isomorphic with real numbers, when all the conventional mathematics we learnt at school and university can be used; in particular, we can cheerfully go ahead and optimize. Whether the concepts of preference and belief as exercised by people do obey such laws is not questioned. Indeed, the blurb for Lindley's latest book describes it as a book about how people ought to behave. How dare anyone not behave according to the laws of my algebra!

The theory is based on the following prescriptions:

People should not have non-transitive preference rankings.

People ought to act as they say they would when faced with the actual as opposed to a hypothetical situation.

People ought to believe that the rules of logic should dicate their behaviour.

People do not do any of these things. Any theory which is worth using predicts how people will behave, not how they should, so we can do our mathematics.

Perhaps the most damning criticism is that of Rappaport. His parody of the monologue between consultant and client runs thus: "You are asking me if you should do A or B? Now this depends on your preference for X to Y. Do you prefer X to Y?" (Yes/No) "Now if you follow my argument carefully you see that this preference implies if you wish to be consistent (and I am sure you do) you should do A. Thank you, £1000 please. Yes, you are quite right, all I have done is ask you in a round-about way if you prefer A to B. As you prefer A, I recommend you follow your preference." Thus the fee is earned for a completely tautological procedure. It could be argued the fee is earned for showing the client how to be consistent. But why should he be? We want him consistent so we can apply our techniques to his problem. He has an easy way out of the situation. He still takes the path he wants to but simply changes his stated preferences so our analysis shows him consistent. If we are unfair enough to hold him to those preferences in some other project, it only exercises him a little to explain what differences in this new project justify him in changing his preference.

We have all taken part in such games and found them very frustrating; the idea of ossifying them into a formal theory is an act of masochism in which I want no part.

It is freely admitted that both theories are about individuals and that preferences and beliefs can be different between two individuals. But when we face conflicting objectives they are generated by such different preferences. Neither theory tells us how to combine or resolve them, nor more important, how the individuals forced to act in unison will change their preferences and resolve their conflict.

The usual plea against such arguments is that "You don't tell us what to do. We must do something. This, for all its faults, is all we can do." My rejoinder is that I would rather do nothing than something which may lend an air of respectability to a thoroughly unsound judgement. Let us be frank and admit some things are beyond our powers to handle yet, and rely on people's judgement, based on more experience than we have, rather than a shabby trick which makes them think we have done better than they could unaided. These theories try to overthrow those judgements on the basis of other judgements made by the same person, which he understands less well and is less capable of making. It might be more sensible to reverse the whole process and determine the judgements needed on these latter matters to justify his main judgement.

Of course, all the questioning that goes on is valuable to both client and customer and every competent OR worker asks such questions. But they should be aimed at giving an understanding of the problem and no more.

K. D. TOCHER

In his recent paper 'Discussion on Control'[1] Professor Tocher argues that utility theory is based on the naive premiss that people act (or should act) consistently, and as they do not, the theory is useless – or worse. The inconsistency can take the form of (a) not having transitive preferences, (b) not doing in real situations what they say they would do in similar hypothetical situations and/or (c) acting illogically. No doubt Toch bases his views on his observations of people with whom he comes into contact. I can only wonder what sort of people these are since my own impression is that most of the people I meet, professionally and otherwise, are perfectly consistent *most* of the time, that is to say I can form a pretty good picture of what they like and what they don't like; I can rely on what they say and, unless acting in a rush or panic, they behave generally in what seems to me to be a rational and logical way. I would indeed suggest that if it were otherwise we would either have grown up with different views of what is rational, or our institutions and perhaps ourselves would have ceased to exist. Granted there are exceptions, but these can usually be traced to a relatively small number of reasons. In certain circumstances, for example, a person may deliberately tell lies. On the other hand, the complexity of a situation may be such that an individual is unable to examine fully all the logical interactions with the result that he may act in such a way that he would later on see to have been in error – this is particularly likely to happen when decisions have to be taken in a rush, of course.

Toch's "apple and orange" example can hardly be used to damn utility theory, since the theory nowhere states that preferences cannot change over time, or with the state of the individual. Furthermore recent developments in the theory extend it to allow for the fact that people's discriminatory powers are finite and that they may be uncertain whether they prefer A to B, etc. Toch is surely aware of this. However, since in most purported applications of utility ideas the preferences which should be ranked are usually for sums of money which already have a natural ranking, I would be very surprised to find non-transitive preferences – indeed I should be very concerned if I were a shareholder in a company in which I discovered that management had non-transitive preferences for money (all other things being equal of course).

If Toch is right and it is not possible for management to determine preferences and communicate them to us we might as well shut up shop. Whether Rappaport accepts it or not O.R's contribution to man's welfare, for which we are paid, has been to point out that if you take action X this leads to outcome A, and if you take action Y this leads to outcome B – and if you prefer A to B then you should do X rather than Y. One might say that the real O.R. contribution is in the first part of this – deriving the logical relationships between X, Y and A, B. However, if management does not know whether or not A is preferred to B what is the point of knowing the relationship? If, however, management does know that A is better than B, then it is worthwhile for the O.R. analyst to find this out, since it can make for more efficient methods of analysis. Clearly, some means of ranking outcomes must be available if we are ever to be able to say we have made an improvement in anything.

Certainly, utility theory as expounded in some text-books has many short-comings, though they are not the ones that Toch mentions. One problem as Borch[2] points out is the static nature of the theory. Another might be due to the fact that the questions we are supposed to ask the decision maker are ill-defined and ambiguous. For example, text-books often claim that one can draw up a utility function of present value or DCF yield in project appraisal. However, if, as I have argued elsewhere[3,4], these indices are not "sufficient statistics" to describe the project, it is not surprising that inconsistent answers are sometimes obtained. Furthermore, decisions typically interact with one another, so it is important to specify the environment carefully when exploring a particular decision problem. If this is left ambiguous, then again the analyst must expect to reap the consequences.

Toch has, of course, a well-earned reputation for the originality and depth of his thought. The naive person might think that he was just practising what he preaches about man's inconsistency in some aspects of this paper. A deeper analysis will show, however, that he is simply he is simply indulging his penchant for teasing and that he is in fact being consistent after all.

O.R. Department   R. M. ADELSON
University of Lancaster

REFERENCES

[1] K. D. Tocher (1976) Notes for Discussion on "Control". *Opl. Res. Q.* **27**, 231–239.
[2] K. Borch (1967) The Theory of Risk. *JRSS*(B) **29**, 432.
[3] R. M. Adelson (1970) Discounted Cash Flow – Can We Discount It? *J. Bus. Fin.* **2**, 50.
[4] R. M. Adelson (1971) Discounted Cash Flow – The Other Point of View. *Moorgate and Wall Street*, **43**.

DISCUSSION ON CONTROL – REPLY TO A COMMENT ON MY PAPER

Adelson's concern that I should be consistent is very touching and I am grateful to him for finding a way to save me from sin, even though it involves casting me in the role of the court jester.

This insistence on consistency is interesting, because it is a good example of the pervasive nature of our machine age thinking. We must think consistently otherwise, since thought is father to action, we may not act consistently. Unless we act consistently our actions cannot be predicted and we cannot be treated as machines subject to the laws of cause and effect.

If, in fact we were always consistent and always had been then there would be no progress and there would have been no progress (perhaps the cynical would prefer the word change rather than progress). Amongst the small band of pioneers who are striving to find a methodology which can escape from machine age thinking and its dehumanising effect on society, Shackle has emphasised that decision taking (if not null or illusionary) is a creative act on the part of the decision taker. Something new cannot be consistent with the old because it involves different behaviour this time from the last similar occasion.

Thus I would prefer to wallow in sin, in Adelson's eyes, than be condemned to be an automaton – a puppet with illusions of free will manipulated by those who have experimented sufficiently with me to know my utility function.

Of course, Adelson is right that there could have been no progress unless humans had been capable of anticipating each other's reactions to situations as they arise – i.e. we have the paradox that creative, non-machine-like behaviour of humans could not be exploited unless most of the time we had acted like machines.

The crucial point is when is it important to act creatively and when as a machine? The machine like behaviour is appropriate in the ordinary day-to-day activities; the creative behaviour is important when decisions with far reaching consequences are being taken. A machine age methodology is appropriate for studying the unimportant activities and not for the important ones.

Another example of the pervasive values of machine age thinking is my own initial reaction to Ron Adelson's criticism of my "apple and orange" example.

Of course we could allow utility to change over time, indeed be a function of our past experience and no doubt with a little creative thought we could cook up a model of how utility is related to experience and give a descriptive theory of the fruit eater.

My initial reaction is that this is useless as a prescriptive theory because the model of the utility function would have to be verified by experiment before it could be applied. Put in cruder terms, if there are practical (never mind the philosophical) difficulties in measuring utility what hope have we of measuring changes of utility?

My mistake, part of the heritage of machine age thinking, is to suppose that my desire to predict the fruit eater's behaviour is an appropriate one. "The fruit eater believes he exercises free will. I know better; he is just a puppet controlled by his genes, utility function or what you will and I, knowing all about these, can treat him like a machine – the law of cause and effect rules supreme."

The law of cause and effect has served us well for 3 or more centuries and we would be foolish to abandon it now. What we are pleading now is for a recognisation that in some crucial situations, it is not appropriate and a new methodology is required.

Adelson seems to regard illustrations that people do have preferences and indeed must have preferences as a sufficient rebuttal of an attack on utility theory.

I am not denying the value of the concept of preference, but do deny that preferences must obey the rigorous laws of combination supposed of them. His criticism of Rappaport's argument is a good illustration.

There are three components

a   the manager prefers A to B
b   an analysis which shows $X \rightarrow A$, $Y \rightarrow B$
c   a decision to do X rather than Y.

Adelson envisages the situation that the manager asserts a, the OR man asserts b and because they are rational men – i.e. believe in the application of the laws of logic, then c will happen.

Consider a more realistic situation. The manager wants to do Y. He asserts he prefers A to B, the analysis shows the inconsistency of these two statements. What happens? The manager either changes his mind (i.e. one of his preferences) or, and, in my experience this is more likely, he denies the validity of the analysis, throws a few more facts into the situation until the analysis shows he was right after all.

If all this is done seriously, it is valuable because all concerned now understand the situation better. In particular, since the manager's preferences presumably have some rationale behind them, it reveals that neither X nor Y is a very good solution and the process may allow the creation of a new possibility Z. The process starts again.

The analysis is a mechanism to make people think about the problem in a structured way and to enhance the opportunity for a new idea to emerge. We don't need utilities to go through such a process; they are irrelevant.

K. D. TOCHER

Is utility theory useful? Professor Tocher's recent, forceful and emotive criticism[1,2] might easily lead one to doubt its value. To use utility theory, he claims, is to deny the existence of free will, to dehumanize society and to treat people as machines. Since our preferences are undeniably complex in the extreme and varying with time, he argues that any quantitative treatment of them must practically, if not philosophically, fail.

There are, I think, two important, related criticisms of Professor Tocher's argument. First, he has assumed that the objective of any quantitative analysis of a problem is to get a precise, quantitative solution. Far from it. My purpose in analysing a situation quantitatively is to get a better qualitative feeling for the underlying problem. Second, he has assumed that the role of utility theory is to predict the decision-maker's preferences. It is not. The role of utility theory is to enable the decision-maker to express his preferences. Let me enlarge upon these points.

Professor Tocher rightly strives to avoid errors of our machine age thinking. The greatest and most pervasive of these errors is that numbers mean something *per se*. They do not. Man is a qualitative rather than a quantitative beast. Indeed, the words "qualitative" and "quantitative" imply a dichotomy of meaning that does not exist. Man learned to count not because he desired to know the number of sheep in his flock from reasons of idle curiosity, but because he wanted to know whether he had lost any; his primary interest was in qualitative questions containing the comparatives "more", "less" and "the same". Over the generations man has found the need to make finer and finer qualitative comparisons and consequently he has developed more and more refined number systems, higher and higher mathematics and more and more skilful techniques of modelling systems. All the time man was striving towards a better understanding of his environment, a better qualitative feeling for his problems. Yet somewhere along the line the "mythology of the quantitative" has sprung up. Numbers suddenly have a meaning of their own. To describe an environment by a single index has become an ideal. It is not an ideal. It is not the objective to which those of us who model systems should strive. Rather we should be trying to get a feeling for the balance of the various interacting components of a system.

Utility theory, indeed the whole of subjective decision theory, is useful precisely because it enables the decision-maker to get such a qualitative feeling for his problem. By expressing his preferences through utilities and his opinions through subjective probabilities as best he can, he may find the decision which best balances his belief in the likelihood of its possible consequences with his feelings of preference for them. But then his analysis is not over. He should investigate the sensitivity of his choice to changes in his preferences for the possible consequences and to changes in his beliefs as to their occurrence. Thus he gets a feeling for how the balance of his preferences and beliefs is made up. In the light of his understanding he then makes a decision. Decision analysis should not directly lead to a decision, only indirectly through the intermediary of understanding.

To pick up my second criticism of Professor Tocher's remarks, in the analysis described above utility theory gives expression to the decision-maker's preferences; it does not predict them. Perhaps Professor Tocher is labouring under the impression that a decision analyst is employed to make decisions (for then he would need to predict his clients' preferences). He is, however, not contracted for this purpose. Rather he is asked to help his client gain a better insight into the interactions between all the components that go to make up his preferences and opinions. The client should still make the decision.

No doubt Professor Tocher will respond that much of what I have said is a re-expression of his own views. Indeed, his description of the real life analogue of Rappoport's example underpins much of what I have argued above. However, he would deny the value of utility theory in getting a qualitative feeling for the problem, because it demands that preferences must obey certain rigorous laws of combination. I remain unconvinced. First, my preferences, when I consider them, do obey the required rigorous laws. Second, in the sensitivity analysis, which is an integral part of the decision analysis, the constraints imposed by these laws may be softened slightly. Finally, even if the decision-maker does discover that the formalism of utility theory does not give satisfactory expression to his preferences, there he still gains insight into his problem.

Department of Decision Theory,   SIMON FRENCH
Manchester University

REFERENCES

[1] K. D. Tocher (1976) Notes for discussion on "Control". *Opl. Res. Q.* **27**, 231–239.
[2] K. D. Tocher (1977) Reply to Comment on above. *Opl. Res. Q.* **28**, 107–109.

Dr French's mode of argument is most interesting; it might be described as reverse-Jesuit. A Jesuit refutation starts with an agreement (with easily accepted qualifications) of the assumptions of the argument and then shows that the qualifications lead logically to exactly the contrary conclusion to that reached in the original argument.

Dr French asserts the contrary conclusion and justifies it by a set of assumptions which he supposes I would agree with. I think the forward Jesuit technique has greater psychological impact.

I do not believe that " the greatest of the errors of machine age thinking" is "that numbers mean something *per se*". This is just a common error of any way of thinking.

French's shepherd learnt to count not only to determine if he had lost sheep, but so he could tell if he had lost more sheep than his rival had gained. The counts give expression to the model.

"Sheep lost = Sheep at first count − Sheep at second count."

I cannot understand how if man were a "qualitative beast" he found it necessary to become more and more quantitative by inventing "refined number systems". Surely the urge to measure arose from the desire to be more precise about the state of nature; this carries with it the need to be more specific about what one is being precise about. A qualitative (vague) statement is used in preference to a quantitative (precise) statement when greater generality is required.

The danger from quantification arises when it gives a false sense of precision, when the law of behaviour invoked does not have the sanctity of experimental verification. A Yahoo shepherd who counts one, two, many is less precise about his flock than Dr French would be.

We all know examples where studies were made by quantitative models which lead to results of qualitative value. The most famous is queue theory from which an understanding about the unstable nature of balanced systems came from mathematical analysis of simplified queueing systems. However, the original intention was to be precise about specific situations and the final outcome, a qualitative statement about a general situation, was an accident. That is a far cry from Dr French's proposal that we invent the elaborate structure of utility theory to give us insight about the nature of the balance of interacting components of a system.

This argument seems to say that man best understands a situation through examples and utility concepts allow some calculation to be done on some hypothetical examples and then by intuition and a study of several such examples, insight is obtained; that is undoubtedly true for many men. The danger lies in the common hidden assumptions of all the examples – the validity of the rules of combination of the utilities. Examples for which these rules do not apply are never considered. Now either the assumptions are used when the validity of the conclusion is suspect or they are not when the theory is unnecessary.

Thus the prescription seems to be: come to a vague conclusion based on a set of unvalidated precise examples. I suppose it is common ground that in many situations all that can be done is to come to a vague conclusion. What I want is a precise logically valid method of reaching that vague conclusion. Utility theory gives no mechanism for reaching that conclusion from the examples – that is left to intuition.

A far more promising approach is through fuzzy set theory which enables valid statements to be made about the implications of imprecise statements. This is a theory which tries to replace the intuition by a precise form of argument.

However, I think Dr French would be compelled to admit that his view of how to use utility theory is a very sophisticated one, not commonly held by most exponents of that theory, certainly not expressed publicly.

Turning to the second criticism by Dr French, he charges me with usurping the decision-maker's role and not sitting meekly at his side. This is a hoary old discussion point and is best left as a statement that analyst and decision-maker working together reach some conclusions which lead the decision-maker to executive action.

There is some ambiguity in the criticism "utility gives expression to the decision-maker's preferences but does not predict them". It is true that the decision-maker supplies those preferences the analyst requires for his calculation; this calculation then predicts what the decision-maker should prefer to do. If the analyst has persuaded a decision-maker to play the game with him, then the decision-maker will behave as predicted unless he takes further factors into account or ignores his advisor for some other reason. If the decision-maker does not believe the analysis then he probably would not play anyhow; if he does not understand the analysis then he is unlikely to obtain any insight from the study. If he takes further factors into account, then the analyst should enquire about a few more preferences and present a new analysis. I cannot see any merit in getting the worst of both worlds by combining an intuitive guess of the effect of a new factor with a precise analysis of the rest.

Of course, in many important situations the decision-maker is a myth. There is no one person who determines the future course of events; the efforts of one to obtain his goals however well he understands them, is frustrated by others with differing objectives. The situation is game-like and although classical game theory is riddled with paradoxes and is too rigid to describe a real political-game it does model the essential features of many situations in which executive choice is required.

Any theory of decision-making that fails to recognize the existence of an element of conflict between parties that must be resolved, cannot be regarded as realistic. Utility theory only recognizes conflict within an individual and not between parties of people.

This introverted aspect of utility theory is a reflection of the essentially idealistic philosophy underlying it; ideas transcend reality. Dr French reveals this when he asserts that his preferences always do obey the combination laws when he considers them, i.e. he adjusts his preferences by considering them to ensure that his assertions are self-consistent.

Now any analysis of a conflict situation must allow for not merely people's beliefs about what is true and their desires for what will be true, but also their hopes for what can be made true and these depend on the beliefs, desires and hopes of the other protagonists. The analysis cannot use these but must depend on the belief of the one player of what these beliefs, desires and hopes are for the others. Each player must calculate the predicted behaviour of the others before he can choose his own action and these predictions depend on his own choice. This recursive feature must be broken by some mechanism.

Thus, in my opinion, an analyst is concerned, when guiding his client, with predicting the behaviour (i.e. the decisions) of others and this analysis assumes that he can predict his client's behaviour at least as well as the opponents' analysts.

I am always grateful to people who leap to defend utility theory because every defence betters my understanding of its irrelevance to modern society.

K. D. TOCHER

Professor Tocher continues to challenge the relevance of utility theory to modern society.[1-3] His early attacks[1] were based upon the argument that the interactions of a man's beliefs and preferences are simply too complex to be usefully represented by the quantitative concepts of subjective probability and utility. Later, in his replies to defences by Adelson[4] and French,[5] he has developed his argument in various ways. It is upon his latest reply[3] to French that we comment below.

Professor Tocher's initial remarks on the counting habits of shepherds and on the objectives and results of research into queueing theory seem, if anything, to support rather than dismiss French's argument. However, we do disagree on one point. We do not accept that "vague" is a synonym for "qualitative", nor "precise" for "quantitative". To suggest they are is a nice example of the forward Jesuit argument. We pass on.

Decision analysis, indeed any scientific analysis, involves three steps:

(1) A move from the real world to a set of mathematical models (or "hypothetical examples" in Professor Tocher's terminology).
(2) The finding of solutions to these models.
(3) A move from these abstract solutions back to physical operations in the real world.

We leave steps (1) and (3) entirely to the intuition, simply because we know of no better of man's faculties. Intuition or understanding – call it what you will – is an essential part of our humanity, and we do not seek to deny it a place in our decision-making. Indeed, we believe it to be irreplaceable. Professor Tocher may search for a replacement if he so wishes, but we would give him one piece of advice: he will not find it in fuzzy set theory.

He suggests that "fuzzy set theory ... enables valid statements to be made about the implications of imprecise statements". We do not in any sense believe this to be so. The problem with fuzzy sets is that, to our knowledge, no-one has yet given an axiomatization of the set membership function. It may not be a subjective probability distribution or a utility function or even a combination of these, but it is certainly a numerical representation of a weak order. Measurement theory[6,7] tells us that such representations are based upon assumptions of the same nature as those underpinning utility theory: those assumptions whose validity Professor Tocher repeatedly denies.

In the last analysis fuzzy set theory is based upon ordinary set theory (the set membership function cannot be defined otherwise). Similarly, fuzzy logic is based upon the more usual two-valued logic. Implicit in all fuzzy mathematics are all the assumptions explicit in conventional mathematics. So it is only possible for fuzzy mathematics to resolve problems unamenable to conventional mathematics if it introduces further restrictive assumptions. Where have these assumptions been stated? Utility theorists have been quite open about their assumptions; the proponents of fuzzy mathematics have not.

Returning to French's "sophisticated" and "not commonly held" view of utility theory, he gained it from the literature.[8-10] In particular, the studies of Keeney and de Neufville[11] on the siting of Mexico City Airport and of Keeney and Nair[12] on the siting of nuclear power stations in the State of Washington are indicative of the sensitivity now possible with expected utility theory. The methods employed displayed to the decision-makers the complex interactions of their opinions and preferences allowing an agreed decision to be reached.

It is precisely because we recognize that decision-makers are seldom unique individuals, but rather groups of individuals, that we propose the application of decision analysis to real problems. It is true that the multiple-person decision problems in theory need have no simple solution of the type found for the single decision-maker. Arrow's Impossibility Theorem[13] shows this. But it is a theoretical result that applies to any collection of individuals with any set of preference patterns, however disparate. In practice, a group of decision-makers does have a common core of objectives. Indeed, that is how we recognize them to be a group. For such a group decision analysis brings much understanding. First, the expression of beliefs and preferences in terms of subjective probabilities and utilities enables each member of the group to express his views effectively to the other members. Second, the problems involved in trying to develop possibly non-existent group probabilities and utilities expose any underlying conflicts and bring them openly into discussion. Third and last, a knowledge of the paradoxes of social choice theory[13] brings with it a recognition that there has got to be "give and take". Without the willingness of the participating decision-makers to moderate their own beliefs and preferences in the interests of group stability, the group cannot be said to be behaving rationally. The Mexico airport study provides a particularly striking example of this process.

It is interesting that Professor Tocher uses game theory to illustrate the conflicts between individuals. As is well known,[13] game theory is based upon utility theory. Furthermore, game theory is also based upon the maximin philosophy of pessimism. Does Professor Tocher really believe that pessimism is an integral part of good decision-making? If so, does he think it to be more part of good decision-making than the philosophy of consistency which he takes so many pains to deny?

Department of Decision Theory,   SIMON FRENCH
University of Manchester   ROGER HARTLEY
    LYN THOMAS
    DOUG WHITE

REFERENCES

[1] K. D. Tocher (1976) Notes for discussion on "control" *Opl. Res. Q.* **27**, 231–239.
[2] K. D. Tocher (1977) Viewpoint. *Opl Res. Q.* **28**, 107–109.
[3] K. D. Tocher (1978) Viewpoint. *J. Opl Res. Soc.* **29**, 180–182.
[4] R. M. Adelson (1977) Viewpoint. *Opl Res. Q.* **28**, 106–107.

[5] S. French (1978) Viewpoint. *J. Opl Res. Soc.* **29**, 179–180.

[6] J. Pfanzagl, (1969) *Theory of Measurement*, Physica-Verlag, Würzburg–Wien.

[7] D. Krantz, R. D. Luce, P. Suppes and A. Tversky (1971) *Foundations of Measurement*, Vol. 1, Academic Press, New York.

[8] H. Raiffa and R. L. Keeney (1977) *Multiobjective Decision Making*, John Wiley, New York.

[9] D. J. White (1975) *Decision Methodology*, John Wiley, New York.

[10] G. M. Kaufmann and H. Thomas (Eds.) (1977) *Modern Decision Analysis*. Penguin, Harmondsworth.

[11] R. de Neufville and R. L. Keeney (1972) Systems evaluation through decision analysis: Mexico City Airport. *J. Syst. Eng.* **3**, 34–50. Reprinted in reference 10 and reported in reference 8.

[12] R. L. Keeney and K. Nair (1977) Selecting nuclear power plant sites in Pacific Northwest using decision analysis. In *Conflicting Objectives in Decisions* (D. E. Bell, R. L. Keeney and H. Raiffa, Eds.), John Wiley, New York.

[13] R. D. Luce and H. Raiffa (1958) *Games and Decisions*. John Wiley, New York.

It must be a good thing that utilitarians feel the need to hunt in packs of four. My first reaction to the impressive looking list of references given by French *et al.* was one of despair; first because there were still so many learned gentlemen prepared to defend idealism but more practically because it meant so much reading before I could reply.

However, I buckled down to it and started with the books on measurement. The first of these fell open at a page on which I read, "These . . . are normative principles defining the concept of rational behaviour rather than a description of actual behaviour. . . . We want to stress that subjective probabilities are means of describing rational behaviour. Nothing more! They cannot be used as estimates of the objective probability of an event or the credibility of a statement or for the corroboration of a theory."

With references like this, who needs enemies?

This sums up my attitude to the utilitarians; I am irritated by their arrogance – they will tell me how I ought to think regardless of the evidence of how people actually think or take decisions. They form a band of intellectual Mary Whitehouses who know best for all of us.

This arrogance shows in their writings; I "continue to challenge the relevance" in spite of their putting me right. They give me advice of how to search for something I don't want to find, after explaining the nature of scientific method to us poor ignorants who do not understand it.

This explanation is a good place to start a reply. We all know model building only gives an account of part of any system we study and that the mappings between the real world and the models involves art, intuition – call it what you will – something external to the scientific method. In our case, the crucial issue is what is included in the model and what is left to intuition. My stance is that with our current state of knowledge, the utilitarians try to put too much in the model and do not leave enough to intuition. That is a far cry from "searching for a replacement for intuition". I would rather have no model at all than a bad one. I regard a model of how people think and act which is in conflict with the facts, as a bad model.

Now, of course, to make progress we must try to model the processes we currently leave out. Thus the interest in fuzzy set theory, Shackle's and Ackoff's work and metagame theory. I do not believe any of these are adequate and would not advocate their systematic use in decision-making, but, at least, they are plausible accounts of real behaviour.

The utilitarian critique of fuzzy set theory seems to be that it is fuzzy. They feel that unless it can be explained in terms of old concepts it cannot be trusted. If it could be explained in these old terms it must contain nothing new. All our useful concepts went through a phase when they were not understood and when they had been proved useful, academics then explained them to us. If we would all recognize that we are blundering about trying to understand the mechanisms at work in social decision-taking and not act as if all the answers were known, more progress might be made.

The accounts given of the group decisions about the siting of airports and nuclear power stations emphasizes that a group of people with a common basic objective formed an alliance to drive the projects through. The evidence from the British experience is that these decisions are taken and justified afterwards (with perhaps some minor concessions to the enemy).

It is this element of conflict which is under-emphasized in utility theory and which leads to an interest in a game-like approach. I was not advocating the use of classical game theory which suffers from the weakness of leaning too heavily on utility, but merely of using the basic concept that a decision is taken by acting out or simulating a conflict with all the actions and counter-actions being imagined rather than executed.

Utility theory is concerned with the resolution of internal conflict (in the individual or the group). My concern is with the cause of the internal conflict. I believe that this is just an imaginative anticipation of the external conflicts which are the real cause of the problem.

The complexities of real decision-taking arise from the knowledge that this simulation of the conflict will not be carried out perfectly and that mistakes will be made by the parties to the conflict. This can be exploited to deliberately mislead one party to believe one action would be taken in a given set of circumstances when in fact another action would be taken. Bluffing and all the other political tricks obscure the prediction of the future consequences of a decision.

We have all experienced negotiations in which the objective has gradually changed from achieving some particular result to winning and then to ensuring that the other side does not win, even at the expense of not winning either. A principle of industrial relations negotiation is that the compromise reached allows each party to feel that they have not lost – it has very little to do with the original dispute.

Utility theory seems naive about these realities of decision-making and builds a model of human behaviour which is more optimistic about innate good intentions than I am prepared to accept.

British Steel Corporation   K. D. TOCHER

Tocher, throughout the debate, was firm to his belief, stated so clearly in the first extract:

> any theory which is worth using predicts how people will behave, not how they should, so we can do our mathematics.

He believed deeply and fundamentally that his role as an operational research scientist was to observe, describe and model, but never to interfere. For him, although undoubtedly there was a distinction between *is* and *ought*, it was irrelevant. It did not matter how people ought to behave; what mattered was how they did. For a decision analyst the situation, if not quite the reverse, is certainly very different.

Taking such a very different view, as I do, from Tocher, it is unfortunately very easy for me to give the impression that I do not respect his opinion. Far from it, I do. He was one of the great founding fathers of operational research. During our exchange of 'Viewpoints', a time when I was just beginning in operational research, he and I corresponded several times. Although he disagreed with my then emergent opinions, he encouraged me to develop my ideas. Indeed, that correspondence and encouragement, in part, has catalysed and influenced much of my later work.

4.2   WHAT IS A NORMATIVE ANALYSIS?

Decision analysts have a standard counterargument to the accusation that utility functions and subjective probability distributions are inapplicable because they do not provide valid models of decision makers' actual preferences and beliefs – and lest it be thought otherwise, let it be admitted that we all, proponent and opponent, agree that they do not (Hogarth, 1980; Kahneman *et al.*, 1982; Wright, 1984). The counterargument is that the intention is not to describe the decision makers' preferences and beliefs as they are; it is to suggest what they ought to be, if the decision makers wish to be consistent. One must not confuse 'is' with 'ought', and decision analysis suggests how people ought to choose. Decision analysis is an example of a **normative** (or **prescriptive**), not a descriptive, analysis.

Although I accept the spirit of this counterargument, I do concede that the bald manner in which it is usually presented is not entirely conducive to the winning of the debate. It seems to grant decision analyses almost dictatorial powers over our behaviour. Yet those of us who believe in the value of decision analysis do not see it as providing a dictatorial strait-jacket of rationality. Rather, it is a delicate, interactive, exploratory tool, the purpose of which is to help decision makers understand their situation, beliefs and preferences better so that through this understanding they may make a more informed decision. It is this latter, softer view of decision analysis that I wish to explore in this section.

The problem with claiming that a decision analysis is an example of a normative, not a descriptive, analysis is that the distinction between normative and descriptive analyses has not been explored as much as it might have been in the literature. Usually it is stated simply that a normative analysis tells us how we ought to behave in particular circumstances, whereas a descriptive analysis conjectures how things are behaving. The methodology of descriptive analyses has been well discussed. It is usually referred to as the scientific method; its application within operational research is considered in detail by, *inter alia*, White (1976). The methodology of normative analyses has been less well discussed and it is there that I shall concentrate my attentions. For obvious reasons I shall only consider normative decision analyses.

To begin the discussion, I shall confine attention to the case of a decision faced by a single person: I shall address the case in which a group is responsible for a decision towards the end of this section. There is some advantage in assuming that I am the decision maker. I shall shortly introduce a model decision maker to whom I shall refer in the third person, thus creating a natural contrast with the actual decision maker – me.

Suppose than that I face a decision: personal or business, it does not matter which, but a complex decision and one that matters to me. Decision problems seldom arrive in a neat, cut-and-dried form with clear alterna-tives. Typically, I would be aware that I face a choice; but I would be unclear as to precisely what that choice is and, moreover, I would be unclear about my preferences and beliefs in those aspects of the choice that I can perceive. Note that, in suggesting this, I am already implying that a decision analysis cannot be descriptive in nature. It cannot describe my preferences and beliefs as they exist, quite simply because they do not exist yet. I need to think deeply about my problem before I can determine what they are. The purpose of the analysis is to help me think deeply.

It may be advisable to pause and remark that I would be unwise just to sit and ponder without structuring my thoughts through some decision analysis. Many have argued that the human mind has great subtlety. With sufficient thought it can appreciate facets of a problem that any analysis, with inevitable simplifying assumptions, would miss. Decision making is necessarily a *Gestalt* process and it is unwise to supplant holistic judgement. Alas, persuasive as this argument may be in the abstract, it is remarkably lacking in empirical support. Unguided human judgement is susceptible to many failings. I cite some of the evidence for this in my paper

'From decision theory to decision analysis' (French, 1984), reprinted in the next section. Suffice it to say here, that, having read many studies in the psychology literature, I am not prepared to rely upon my holistic judgement.

The first phase of the analysis is to formulate my problem, to identify my options and their possible consequences. This means I must construct a decision table or tree representation of my choice: I shall assume the latter. Problem formulation is an iterative process in which I repeatedly ask myself: what can I do; what factors, beyond my control, might affect the consequences of my actions; how might I modify my actions in the light of these factors? Each answer to these questions prompts me to draw a part of the tree. Each part that I draw may prompt a new question. Drawing one part of the tree may make me realize that another part needs substantial revision. It is important to realize that the construction of the decision tree is a creative exercise: it does not simply capture my perception of the choice facing me; it helps form that perception. Wells (1982) has reported that users find the formulation of problems as decision trees: 'a refreshingly clear way of understanding the issues'.

Note that the decision tree is a 'small world' representation of my problem (Phillips, 1982 – reprinted in section 3.6; Savage, 1972). True, it is a representation that is central to forming my perception of that problem: but it is not the problem itself. To emphasize this, I shall refer to it as a **model decision problem**. Necessarily, the model decision problem omits many side issues that are irrelevant to the main choices before me. For instance, in the study by Phillips (section 3.6) there are no branches of the tree corresponding to the choice of colour of the engines. Yet that is a decision that would have to be made; but, of course, it is a decision of peripheral importance compared with the main choices of whether to develop microchip technology, etc.

The decision tree at this point of the analysis will have the consequences represented holistically. The next stage is to represent these as vectors of attributes. Developing such representations is a reflective procedure, which again helps form my perception of the issues before me. When this is done, I turn my attention to my preferences and beliefs.

Remember that I am assuming that, initially, I am muddled and confused about the problem before me. In particular, I am unsure of my preferences and beliefs. Developing the decision tree will have clarified the main issues, and seeing those more clearly may have made me a little surer of my feelings about the problem. But in any major, complex choice I would still expect to be unsure to a large degree about what my preferences are and what my beliefs are. Perhaps it would be better to say that I am unsure about how to balance my preferences for conflicting objectives, about what to believe given the many possible ways that future un-certainties may be resolved, and, moreover, about how to balance my preferences with the inherent uncertainty in the problem. In short, my thoughts and feelings about the problem need 'straightening out'. Assessing utilities and probabilities helps me do this.

Consider the assessment of utilities. To straighten out my preferences, I first need to consider exactly what I mean by 'straight'. In judging which independence/dependence conditions I believe should hold among my preferences for the attributes, I am essentially defining 'straight' for the context of the problem. These conditions define the form of an appropriate multi-attribute utility function, and the process of assessing this guides me towards such 'straight' preferences. I further explore this interpretation of the assessment process in French (1983b, 1986).

Similarly, the assessment of subjective probabilities helps me sort out and 'straighten' my beliefs.

Note that both assessment procedures are reflective and constructive. They use the framework provided by the independence/dependence conditions, which I have identified as appropriate, to structure the assessment. I am asked simple questions and from the answers to these the framework shows me the implied answers to more complex questions. If these seem sensible, I accept them. If one causes me some discomfort, then I reflect upon the relevant aspects of the problem until I realize that the framework is wrong and needs modifying, or that one or more of my answers to the simple questions do not truly reflect my feelings, or that I was wrong to be discomforted by the implied answer to the complex question. Whatever the case, I gain insight; and my preferences and beliefs evolve.

Next, the analysis calculates the expected utilities of the branches in the tree. The ranking provided by these suggests how I might balance my preferences and beliefs to identify a 'best' course of action. I find it helpful to think of this stage of the analysis in the following way.

I believe that a normative decision analysis guides my choice, in part, by example. First, it constructs a model decision problem: that we have seen. Next, in the context of the model decision problem the analysis erects a **model decision maker**, who in one sense is idealized and in another reflects me. The utility function and subjective probability distribution are this model decision maker. They give his beliefs and preferences.

He is idealized in that his logical structure is always consistent with certain axioms, e.g. transitivity of preference, which I take to embody canons of rationality. Thus he is idealized in a way that I admire: the consistency inherent in his preferences and beliefs is one that I would wish to emulate. Subject to this constraint of consistency, the model decision maker reflects me in that his preferences and beliefs model my own. If I prefer more money to less, then so does he; if I am risk averse, then so is he; if I believe that it is more likely to rain tommorrow than snow, then so does he: etc. At least, he reflects my judgements providing that they are reasonably consistent with the canons of rationality. If they are not, I must either modify them or reject the whole analysis, since then I clearly do not wish to emulate the behaviour embodied in the model decision maker.

Using the logical structure of the model decision maker, I can now determine his choice in the model decision problem: he will choose an action that maximizes his expected utility. Thus I can see how an idealized decision maker with preferences and beliefs similar to mine would choose in a situation, the model decision problem, which parallels my own. Because I admire the idealization present in him, his choice will be a pointer – but only a pointer – to my choice in my real decision problem.

In the above, I have been referring to the model decision maker, as if he were uniquely defined. Of course, he is not. The assessment procedures only locate my utility function and subjective probability distribution to within certain limits. Remember that in sections 2.2 and 2.3 I admitted that none of us could state the precise indifferences that the procedures apparently require: our powers of discrimination are finite. Thus there is no single model decision maker to provide a uniquely correct example for me to ponder. Rather there is a family of such decision makers, all of whom provide appropriate examples. Sensitivity analysis can be seen as the process that enables me to examine the choices of this family.

The output of a decision analysis is not a prescription of which action to take: it is understanding. The analysis is meant to bring an understanding of where the balance of my beliefs and preferences lies. Sensitivity analysis helps me appreciate the sharpness of that balance. If the family of model decision makers all agree, the balance is clearly well defined. If there is disagreement, it is not so well defined. I must think a little more, analyse a little more. But, whatever the case, my understanding is enhanced.

Note that throughout the analysis there is a very different emphasis to that which would be found in descriptive modelling. There is a clear intention to help me evolve my beliefs and preferences, change and refine my perceptions, both of the problem and of myself: in short, to affect that which is modelled.

I should remark perhaps that decision-analytic modelling is seldom as linear as I have described it here. One is forever iterating, doubling back, occasionally jumping forward, so that it is very rare for a decision tree to be completely built before one assesses subjective probabilities and utilities. In turn, 'rough' expected utilities are often calculated and sensitivity analyses conducted before the assessment of probabilities and utilities is complete. In this respect decision-analytic modelling is no different from that anywhere else in operational research (White, 1976).

So far I have concentrated on how a decision analysis can guide a single decision maker. What happens when a group is responsible for the decision?

The first point to note is that many difficulties can arise if we try to think of a group as a single decision-making entity (French, 1986). It is better to think of a group as a social process which translates the decisions – votes – of its members into a course of action. Thus we should ask how a decision analysis can make the social process smoother.

At the lowest level the analysis advises each group member individually: it enhances his perception and understanding precisely as it does in the single decision maker case. But, because the analysis is conducted within the group, that perception and understanding is shared between the members (French, 1986; Phillips, 1984). Moreover, because each member sees the analyses appropriate to the other members, he sees and understands their views better. The analysis facilitates group communication. It can highlight areas where a difference of opinion matters, i.e. points to different choices of action; and avoid heated but sterile debates over quite irrelevant issues, which do not affect the choice of action. Sensitivity analysis plays a vital role in fostering this communication: see Phillips (section 3.6).

4.3 FROM DECISION THEORY TO DECISION ANALYSIS

We shall discuss the role of decision-analytic models in group decision making further in section 4.4. There we discuss decision conferencing and emphasize that practical decision analysis requires many more skills than competence in decision theory. Before turning to that, however, it may help to summarize the theory and arguments that have gone before. The paper of mine reprinted below emphasizes the structure inherent in a decision analysis and the ways in which this structure may provide a framework for thought.

From Decision Theory to Decision Analysis

S. FRENCH
Department of Decision Theory, University of Manchester, England

When it was first developed, Bayesian decision theory provided such an idealised, simplified view of decision making that it was difficult to believe that the theory had anything to offer practice in the complex, ill defined world of real decision making. Gradually it has become apparent that it has. It can offer structure: structure in which to explore and analyse a problem; structure in which the decision makers can articulate their views and communicate with each other; structure in which parts of a problem can be analysed independently of other parts: and, in general, structure in which to organise thought.

INTRODUCTION

In 1954 Savage laid the foundations of modern Bayesian decision theory.[1] His works had been foreshadowed by that of Von Neumann and Morgenstern on game theory,[2] by that of Wald on statistical decision functions,[3] by that of De Finetti on subjective probability,[4] and, in general, by that of Ramsay.[5] Savage's achievement was to show that the preferences and beliefs of an idealised rational decision maker facing a choice between uncertain prospects should be modelled by utilities and subjective probabilities. At least, they should be if you accept Savage's axiomatic definition of rationality. I, for one, do and for the purposes of this paper I shall assume that you do too. Excellent introductions to the foundations of the theory may be found in the books by Lindley[6] and Raiffa,[7] and a recent discussion of the controversy surrounding them in the collection edited by Allais and Hagen.[8]

We shall begin from the subjective expected utility model of an idealised rational decision maker and see how decision analysis has developed from this to help real decision makers structure, understand and resolve their problems. In the next section we refresh our memory of concepts such as subjective probability, utility, decision tables and decision trees. In the following section we discuss how a multi-attributed representation of the consequences of decision problem *may* allow structure to be introduced into the utility function. We note that similar structure *may* be introduced into the subjective probabilities when the states have a multi-parameter representation. Gradually, it will become apparent that *structure* is the "buzz word" of this paper. In the early sections we concentrate on bringing to the fore the structure inherent in Bayesian decision theory. In the later sections of the paper we turn to recent research in psychology, which suggests that real decision makers are typically very poor at structuring their problems. What they need most from a decision aid is help in structuring their thoughts. It is the marrying of this need for structure to Bayesian decision theory's ability to provide a coherent structure that leads to Bayesian decision analysis.

DECISION TABLES AND DECISION TREES

Most decision problems can be represented, albeit with some simplification, as a *decision table*. The idea underlying this is that the consequence of any action is determined not just by the action itself but also by a number of external factors. These external factors are both beyond the control of the decision maker and also unknown to him at the time of the decision. By a *state of nature* or, simply, *state* we shall mean a complete description of these external factors. Thus, if the decision maker knew the state of nature that would actually hold, i.e. if he knew the true values of the external factors, he could predict the consequences of any action with certainty. We shall assume that, although he does not know the true state, he does know what states are possible. For the sake of simplicity we shall assume that there are only a finite number of possibilities: $\theta_1, \theta_2, \ldots, \theta_n$. Similarly, we shall assume

Table 1. The general form of a decision table

		States of nature			
Consequences		θ_1	θ_2	. . . .	θ_n
Actions	a_1	x_{11}	x_{12}	. . . .	x_{1n}
	a_2	x_{21}	x_{22}	. . . .	x_{2n}
	.	.		. . . .	
	.	.	.	. . . .	
	.	.	.	. . . .	
	.	.	.	. . . .	
	a_m	x_{m1}	x_{m2}	. . . .	x_{mn}

that only a finite number of possible actions are available: $a_1, a_2, \ldots, a_m$. We emphasise that these assumptions of finiteness are made purely to simplify the presentations; they are not requirements of the theory. Letting x_{ij} be the consequence of taking actions a_i when θ_j is the true state, we have a decision table: see Table 1. The symbols x_{ij} stand for complete, holistic descriptions of the possible consequences. In very simple problems they might be numbers, perhaps monetary outcomes, but in the vast majority of cases they are complex descriptions, detailing every aspect of the consequences.

We shall assume that the decision maker has preferences between the possible consequences, and we use the notation:

$$x_{ij} \geqslant_X x_{kl} \Leftrightarrow \text{the decision maker holds the consequence } x_{ij} \text{ to be no less valuable than the consequence } x_{kl}$$

Similarly, we shall assume that the decision maker has beliefs about the relative likelihood of the states, and we use the notation:

$$\theta_j \geqslant_\theta \theta_l \Leftrightarrow \text{the decision maker holds the state } \theta_j \text{ to be no less likely than the state } \theta_l.$$

The problem that faces the decision maker is to construct a ranking, an order of merit, of the actions, based upon $\geqslant_X$ and $\geqslant_\theta$, where

$$a_i \geqslant_A a_k \Leftrightarrow \text{on balancing the inherent uncertainty with his peferences,} \\ \text{the decision maker holds action } a_i \text{ to be no worse than} \\ \text{action } a_k.$$

How should he do this in a rational and consistent manner?

The Bayesians have provided an answer. They have defined properties of consistency between the three relations $\geqslant_X$, $\geqslant_\theta$ and $\geqslant_A$, which should be obeyed if the decision maker is to be considered rational. Moreover, they have made the controversial assumption that the decision maker is willing to extend his problem to include hypothetical gambles with known probabilities. Their threefold conclusion is

(i) a rational decision maker can represent his preferences over the consequences by a *utility function*, $u(\cdot)$, such that

$$u(x_{ij}) \geqslant u(x_{kl}) \Leftrightarrow x_{ij} \geqslant_X x_{kl};$$

(ii) he can represent his beliefs by a *subjective probability distribution*, $P(\cdot)$, over the states such that

$$P(\theta_j) \geqslant P(\theta_l) \Leftrightarrow \theta_j \geqslant_\theta \theta_i;$$

(iii) his ranking of the actions is represented by their expected utilities,

$$\sum_{j=1}^{n} P(\theta_j) u(x_{ij}) \geqslant \sum_{j-1}^{n} P(\theta_j) u(x_{kj}) \Leftrightarrow a_i \geqslant_A a_k. \tag{1}$$

We shall not justify this conclusion in the following, rather we take it as our starting point.

The utility function $u(\cdot)$ and subjective probability distribution $P(\cdot)$ are assessed by asking the decision maker for his preferences in hypothetical gambles. For instance, if the decision maker is indifferent between the two gambles shown in Table 2, it is a simple matter to deduce that $P(\theta_j) = 0.65$.

Unlike probability, which is always scaled so that $\sum_{j=1}^{n} P(\theta_j) = 1$, utility functions are not measured on a fixed scale; there is always freedom to choose the origin and unit of measurement. One convention is the following. Let x^* and x_* be the best and worst consequences respectively in the decision table, i.e.

$$x^* \geqslant_X x_{ij} \geqslant_X x_* \text{ for all } i, j$$

Then set $u(x^*) = 1$ and $u(x_*) = 0$ to define the unit and origin. Having done this it is a simple matter to assess $u(x_{ij})$ for any x_{ij}. Consider, for example, the two gambles shown in Table 3. If the decision maker is indifferent between them for $p = 0.45$

Table 2. An assessment of a subjective probability

Gamble A	Gamble B
£100 with probability 0·65 £0 with probability 0·35	£100 if θ_j turns out to be the true state £0 otherwise

Table 3. An assessment of a utility

Gamble C	Gamble D
x_{ij} for certain	x^* with probability p x_* with probability $(1 - p)$

(which may be determined by adjusting the value of p until he is), then the expected utility representation (1) implies

$$u(x_{ij}) = 0.45u(x^*) + 0.55u(x_*)$$
$$= 0.45$$

We shall not discuss the assessment of subjective probabilities and utilities further. Suffice it to say that there are a multitude of methods based upon similar comparisons of gambles to those illustrated here.[9,10,11] There are two points to note, however, and to which we shall return. First, the structure of Bayesian decision theory separates the assessment of subjective probabilities from the assessment of utilities; beliefs are separated from preferences. Second, the assessment methods include much consistency checking. The decision maker is checked to see if his preferences and beliefs are consistent with the properties assumed of $\geqslant_X$ and $\geqslant_\theta$ respectively. By drawing to his attention any inconsistency, and by allowing him to reflect upon and rectify them, he is guided towards the rationality of behaviour implicit in the Bayesian approach.

The decision table representation of a decision problem is static. It pretends that there is only one point of choice. Yet in real life decisions are dynamic: one decision leads to another, and that to another, etc. For this reason it is more useful usually to represent a decision problem as a *decision tree*. Figure 1 gives the decision tree representation of the decision problem implicit in Table 1. The square at the left of the tree represents the decision maker's choice between $a_1, a_2, \ldots, a_m$, each branch

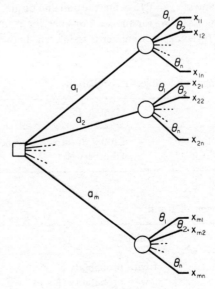

Figure 1 A decision tree representation of Table 1.

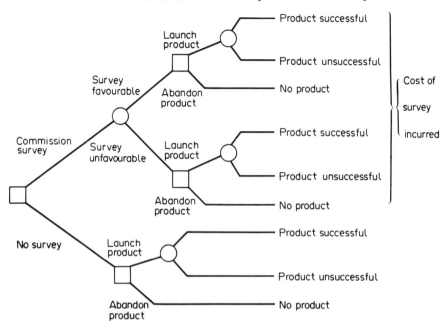

Figure 2 A decision tree of a product launch.

of the tree stemming from this point representing an action, a possible choice. Just as the square represents a decision point, so the circles represent chance points. At the end of each action's branch is a circle representing the uncertainty about the state of nature, and the branch further subdivides into n sub-branches, one for each state. At the end of these is the ultimate consequence x_{ij}. Given the decision tree format, it is a simple matter to represent subsequent decisions by introducing further decision points later in the tree. Moreover, the possibility that certain choices may only be relevant if certain states occur is easily represented by introducing decision points only into the relevant branches. Figure 2 provides a decision tree representation of a decision to launch a new product when there is a possibility of commissioning a market survey before making an irrevocable decision.

Since elementary introductions to decision trees are provided in most operational research and management science courses, we shall discuss them little further here. There are a number of excellent introductions readily available in the literature.[7,12,13] However, two brief points should be noted. Firstly, mathematicians will happily tell you that decision tree and decision tables are equivalent representations of decision problems. Trees can be converted to tables and vice versa: one is the *normal form* and the other the *extensive form*. Decision makers disagree. For them the decision tree representation is far superior: it shows them the structure of their problem. Second, since beliefs are represented by probabilities, it is natural and, in terms of Savage's theory, correct to update beliefs through Bayes' theorem – hence the name Bayesian decision theory. Thus Bayes' theorem places a coherent structure on the probabilities in a decision tree.

THE STRUCTURING OF CONSEQUENCES AS A VECTOR OF ATTRIBUTES

In the above we have represented the consequence of an action by a single symbol, x. Moreover, we have remarked that x stands for a holistic description of the consequence; it is meant to evoke in the decision maker's mind a complete picture of what may occur. And, indeed, in the simplest of cases x may be understood precisely as this. However, in the majority of cases the decision maker may feel, justifiably as we shall see, that carrying a full picture of each possible consequence in his mind is not only unnecessary but may also be confusing in that it clouds the issue with many irrelevancies. In such cases it is better to describe the consequence in terms of a few summary statistics, or *attribute levels* as they are known in decision analysis. The idea is that the attribute levels measure the degree of success or achievement of the consequence against those objectives or factors which the decision maker considers to be the prime determinants of his preference.

Perhaps the easiest way to appreciate what this structuring of consequences as a vector of attributes really is and also how it might be developed is to consider an example. Suppose that a firm is considering where to site a new warehouse. The consequence of the decision will depend on several things: the site chosen, the design of warehouse chosen, the economic situation that transpires and the level of demand that it brings. How should each possible consequence be represented? Upon what criteria does the firm base its judgements of success or failure? Summarised very, very briefly, discussion in the firm might proceed as follows:

What is the prime consideration in choosing between alternatives?
– Maximising cost-effectiveness.
Yes, but what are the prime determinants of cost-effectiveness?
– Financial, temporal and social factors.
What financial factors?
– Construction costs and annual running costs.
What temporal factors?
– Time to build the warehouse, because the launch of a new product may be delayed until its completion; and working life of the warehouse.
What social factors?
– Effect on local traffic congestion and provision of employment opportunities.

Thus the concept of cost-effectiveness is being analysed as in the hierarchy shown in Fig. 3. Indeed, since cost-effectiveness is a meaningless term *in vacuo*, we may say that this discussion is defining the term appropriately for this context.

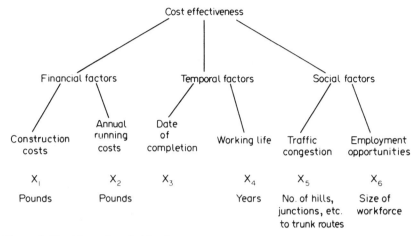

Figure 3 The hierarchy of objectives.

The firm has some way to go to complete this analysis. First, it must discuss deeply and carefully whether it has identified all the factors of importance to the decision. We shall assume that it has. Second, it must decide how to measure the level of achievement of each possible consequence against the six criteria. Possible methods are suggested in Fig. 3. Since the effect of heavy lorries on traffic congestion is most pronounced when they have to change speed, it is suggested that X_5 is measured in terms of the number of road junctions, pedestrian crossings, steep hills, etc. between the warehouse site and the nearest trunk route. This method of quantifying a fairly nebulous qualitative effect by means of a *proxy attribute*, which is expected to be highly correlated with the effect, is common in decision analysis; see Keeney.[14]

This example, despite its brevity, illustrates an important point. The process by which the attribute dimensions are generated is typically hierarchical. A global and loosely stated objective is gradually analysed into more and more detailed sub-objectives. The attributes are the dimensions along which the sub-objectives at the bottom of the hierarchy are measured. The literature contains many exemplary discussions on the methodology of generating such hierarchies so that they are appropriate to the problem, i.e. so that the attributes summarise the consequences in all ways that are important to the decision maker's preferences.[10,14,15,16] We shall say little further here. However, it should be noted that this structuring of the consequences as vectors of attribute levels is not unique to Bayesian decision theory; it is central to many other approaches to multi-objective decision making.[17,18]

Thus we shall henceforth represent consequences as vectors of attribute levels $x = (x_1, x_2, \ldots, x_q)$. Since this a tutorial paper we shall often take advantage of the simplicity of presentation afforded by taking $q = 2$ and write $x = (x, y)$ rather than (x_1, x_2). But it should be emphasised that all the concepts and ideas in two dimensions generalise naturally and straightforwardly to higher dimensions.

Structuring consequences in this way allows the modelling of a very common feature in preferences. Consider the following almost universally agreed statements.

"All other things being equal, I prefer more money to less."
"All other things being equal, I prefer a greater level of safety to a lesser level."
"All other things being equal, I prefer a greater market share to a smaller one."

In terms of two attributes (x, y) these statements correspond to: if attribute y is held constant then a higher level of x is always preferred to a lower level *whatever the constant level of y*. In such circumstances we say that x is *preferentially independent* of y. In general, for q attributes "all other things being equal" translates as: if a subset of attributes are held fixed, then preferences between consequences depend only on the other attributes and are independent of the levels of the fixed subset.

It should not be thought that preferential independence always holds. Let x be the choice of wine with a meal and y the choice of main course. Then convention dictates that red wine is preferred to white if the main course is beef, but that white is preferred to red if the main course is fish. None the less, in many decision problems that do not involve matters of taste, i.e. the majority of problems that concern O.R. scientists, preferential independence does hold.[10,19]

If every subset of attributes is preferentially independent of the remaining attributes, then under mild conditions it may be shown that the utility function must have the form:

$$u(x_1, x_2, \ldots, x_q) = f(v_1(x_1) + v_2(x_2) + \cdots v_q(x_q)) \tag{2}$$

where $v_1(\cdot)$, $v_2(\cdot), \ldots, v_q(\cdot)$ are single dimensional functions and $f(\cdot)$ is strictly increasing.[10]

Preferential independence is a condition which pays no regard to the presence of uncertainty; it concerns preferences between certain consequences. *Utility independence* explicitly considers preferences between uncertain prospects, and in doing so allows further structuring of the utility function over and above that of equation (2). In two dimensions attribute x is utility independent of attribute y if the decision maker's preferences between gambles involving consequences with varying levels of x but a fixed, common level of y are independent of that fixed, common level of y. For example, consider the following four gambles in which the attributes (x, y) are respectively financial prizes received this year and next year.

Gamble E: (400, 300) with probability p; (100, 300) with probability $(1 - p)$.
Gamble F: (250, 300) with probability q; (150, 300) with probability $(1 - q)$.
Gamble G: (400, 100) with probability p; (100, 100) with probability $(1 - p)$.
Gamble H: (250, 100) with probability q; (150, 100) with probability $(1 - q)$.

Notice that in gambles E and F, y is fixed at the common level 300; there is no uncertainty about y implicit in the choice between E and F. Similarly there is no uncertainty about y in the choice between G and H; whatever choice is made and whatever consequence results, $y = 100$. Utility independence requires that the decision maker prefers E to F if and only if he prefers G to H, because the uncertainty in x inherent in both choices is the same.

A decision maker's preferences need not always satisfy utility independence. For example, consider again the situation above in which the consequences are financial prizes in each of the two years. But now consider the following four gambles.

Gamble I: (300, 100) with probability 0·5; (300, 500) with probability 0·5.
Gamble J: (300, 280) with probability 0·5; (300, 270) with probability 0·5.
Gamble K: (25000, 100) with probability 0·5; (25000, 500) with probability 0·5.
Gamble L: (25000, 280) with probability 0·5; (25000, 270) with probability 0·5.

In this case if y were utility independent of x the decision maker would prefer gamble I to J if and only if he preferred K to L. But many would prefer gamble J to I because there is less risk in J even though gamble I has the higher expected payoff in the second year. However, in the choice between gambles K and L many would prefer K to L because, although it is more risky, the certainty of receiving £25000 in the first year means that they can afford to take the risk. In general, two attributes are utility independent if the decision maker's attitude to risk in one is independent of the level of the other.

If two attributes are utility independent, it may be shown that the utility function must have the functional form

$$u(x, y) = u_1(x) + u_2(y) + ku_1(x)u_2(y) \tag{3}$$

where $u_1(x)$ and $u_2(y)$ are single dimensional utility functions and k is a constant.

We have indicated two types of independence assumption that might be appropriate to a decision maker's preferences. There are many other possible types of independence that might hold.[10,20,21] Moreover, there are also conditions that recognise certain types of dependency between preferences for the different attributes,[22,23] particularly the types of dependency that may arise when the attributes have impacts at different times[10,24] (N.B. preferences between gambles I, J, K, L above may fail utility independence precisely because attribute x is received before attribute y.) Different sets of dependence/independence assumptions lead to different functional forms of the utility function; but all these forms have one feature in common: $u(x_1, x_2, \ldots, x_q)$, a function of q variables, is constructed in a simple way from functions of many fewer variables, often functions of a single variable. This structuring of $u(x_1, x_2, \ldots, x_q)$ has important implications for its assessment.

The method of assessing utilities typified by the gamble comparison in Table 3 requires much of the decision maker's judgement. Firstly, he is required to make

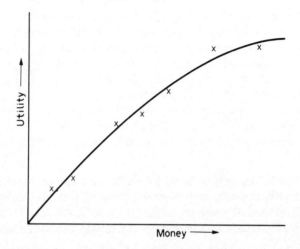

Figure 4 A fitted utility function. N.B. qualitative knowledge of the decision maker's attitude to risk may put constraints on the shape of the utility function. Here he has been assumed to be risk averse[10,19] and so a concave function has been fitted.

one such judgement for each of the possible consequences in the decision table. If the decision table is large – remember it may be infinite – this may require a prohibitively large number of judgements. With the consequences represented along q attributes, each consequence may be plotted as a point in q dimensional space. This allows the possibility of assessing the utility at a representative set of points and then fitting a suitable function approximation to these. This is a well practised technique in one dimension, particularly when the single attribute is money: see Figure 4.[10,25] In q dimensions knowledge of the functional form of $u(x_1, x_2, \ldots, x_q)$ gained from determining appropriate dependence/independence conditions for the decision maker and his current problem identifies a suitable class of functional approximations. But more importantly, knowledge of the functional form allows a more structured, consistent and easier assessment of $u(x_1, x_2, \ldots, x_q)$.

Consider the specific example of $u(x_1, x_2)$ given by (3). If the assessment method is based upon gamble comparisons as in Table 3, the decision maker is asked to make judgements in which uncertainty and the consequent valuation of risk is confounded with the difficulty of trading off one attribute against another. The consequences x_{ij}, x^* and x_* – remember each is now two-dimensional – typically differ in both attributes. It is a difficult enough task to trade off one attribute against another consistently under conditions of certainty. Moreover, the difficulty of this task increases greatly as the dimension q increases. Thus the decision maker may not be able to compare the gambles that he is required to in order to assess $u(x_1 x_2)$. However, here the form of $u(x_1, x_2)$ given by (3) helps;

$u_1(x_1)$ is a utility function for x_1 alone, i.e. for consequences in which x_2 assumes a constant, common value and may, therefore, be ignored because of utility independence. Assuming $u_1(x_1)$ by gamble comparisons is a much easier task than assessing $u(x_1, x_2)$: there are no trade-offs to be considered. Similarly, $u_2(x_2)$ may be assessed. It remains to ensure that $u_1(x_1)$ and $u_2(x_2)$ are consistently scaled, and to assess the constant k. Both these tasks may be achieved by asking the decision maker about indifferences between consequences in situations in which uncertainty is absent. Keeney and Raiffa give details.[10] Thus the structuring of $u(x_1, x_2)$ allows the confounded problems of uncertainty and consistent trade-offs between the attributes to be separated.

Of course, all of the above is predicated upon the assumption that it is possible to identify appropriate dependence/independence conditions for a decision maker and a particular context. Without going into details this may be done by asking the decision maker about his preferences between carefully structured simple gambles of the form of gambles E–L.[10]

THE STRUCTURING OF STATES AS VECTORS OF PARAMETERS

In the last section we saw how the representation of holistic consequences as a vector of attributes led to a structuring of the utility function and a consequent simplification of the assessment procedures. Similarly, the states $\theta_1, \theta_2, \ldots, \theta_n$ in Table 1 were to be understood holistically, but there is much to be gained in structuring them as vectors of parameters. Indeed, the majority of developments in Bayesian statistics over the last thirty years may be seen as investigations of the implications of different forms of structure that may be appropriate to the states. We shall not explore any of these in detail, but rather remark upon one or two important areas and refer to the literature.

At the simplest level if the states θ are vectors of parameters one may enquire whether the decision maker's beliefs about some parameters are probabilistically independent of other parameters. If $\theta = (\theta_1, \theta_2)$ and θ_1 and θ_2 are probabilistically independent then $P(\theta_1, \theta_2) = P_1(\theta_1) \times P_2(\theta_2)$: a well known result, indeed, for many the definition of probabilistic independence. However, there are many other forms of independence.[26] Appropriate use of these conditions can lead to subjective probability distributions based upon hierarchical models[27,28] and mixtures of models, culminating, perhaps with most relevance to decision analysis, in the Bayesian forecasting methodology of Harrison and Stevens.[29] Recent work by David,[30] and Diaconis and Freedman[31] promise many further methods for recognising and introducing structure into subjective probability distributions. Smith[32] gives a recent and full survey of all these developments.

THE NEED FOR STRUCTURED DECISION ANALYSIS

Throughout our development we are implicitly assuming that actual decision makers have need of the support and guidance of some form of decision analysis. It would be well to pause and, at least, to refer to evidence that this is so. Otherwise it might seem simpler, more efficient and more acceptable to allow the decision makers to peruse the complete list of alternatives and to make the choice intuitively without recourse to any formal analysis.

Despite our natural inclination to believe in the ability of the human mind to make well considered judgements and decisions, much evidence has been accumulated by many psychologists to make such a belief untenable. It appears that unguided, intuitive decision making is susceptible to many forms of inconsistency.[33,34,35,36] We do not have the space to give this evidence more than the briefest of surveys, we shall simply cite one particularly cogent piece of evidence.

Kahnemann and Tversky have made many studies of intuitive decision making, albeit mainly under "laboratory" conditions. In one test they asked a group of 152 students to imagine that the U.S. was preparing for an epidemic which was expected to kill 600 people. They had to choose between two health programmes to combat the epidemic.

Programme A would save 200 people.
Programme B would give a 1/3 probability of saving all 600 lives and a 2/3 probability that no-one would be saved.

72% of students preferred programme A. In a second test 155 different students were presented with the same situation. However, they were offered the choice between the following programmes.

Programme C would lead to 400 dying.
Programme D would give 1/3 probability that no-one would die and 2/3 probability that 600 would die.

78% of the students preferred programme D. Although two different groups of students were used in these tests they shared a common background, so it is reasonably safe to conclude that generally there is a tendency to prefer A to B and to prefer D to C. Yet a moment's reflection shows that programmes A and C are effectively the same, as are programmes B and D. Thus there is evidence that people's preferences may be dictated by the presentation of a problem and not by its underlying structure.

The moral that we draw is that a decision analysis should help a decision maker explore his problem and come to understand it. It should challenge his view of the problem with other views to ensure that he fully appreciates all aspects.[37] Many other experiments have indicated other flaws in intuitive decision making. Decision makers may:[33-38]

trade off conflicting objectives inconsistently;
base their decisions upon mutually-contradictory beliefs.
assimilate new evidence into their beliefs very poorly;

adopt very poor heuristics to cope with decision making in an uncertain
    environment; and

communicate with other decision makers very poorly.

All this evidence suggests that some form of decision analysis is necessary to help decision makers structure, understand and fully appreciate their problem; they should not rely upon their intuitive abilities. It is my conviction that the logical structure of Bayesian decision theory is so persuasive that a decision analysis should treat it as an ideal and guide decision makers towards such consistency. In the next section we indicate how this may be achieved.

THE STRUCTURE IN A BAYESIAN DECISION ANALYSIS

In the early parts of this paper the structure inherent in Bayesian decision theory was emphasised. Above we have suggested that structure is sadly lacking from intuitive decision making. The practice of decision analysis that has evolved over the last twenty years uses the former to satisfy the latter's need.

First, the decision problem is explored by the decision analyst and decision makers and a decision tree representation of the possible alternative strategies developed. In developing the decision tree the decision makers come to appreciate the choice that faces them. Indeed it may clarify the choice before them so much that they are able to make the decision without further analysis because one alternative is so clearly superior.[39] (This last observation may not in itself be a triumph of Bayesian decision analysis, but it does emphasise that, if decision makers are left to their own devices, they may even fail to appreciate the true alternatives that face them.)

Representing the consequences as vectors of attribute levels ensures that they are compared with each other along the same set of dimensions. Many have argued[18] that no multi-attribute formulation can capture the subtlety and delicacy of the human mind's ability to compare holistic alternatives. Persuasive as this argument is in the abstract, it has an unfortunate lack of empirical support. Holistic human judgement tends to give weight to fewer attributes than a guided multi-attribute approach.[40] Moreover, different attributes may be considered in different comparisons. Using vectors of attributes to represent consequences counters many of these vagaries of holistic judgement; and the identification and enforcement of an appropriate set of dependence/independence conditions is a further insurance, since appropriate and consistent weighting of each dimension is ensured. Furthermore, the whole process of assessing an appropriate multi-attribute utility function helps the decision maker explore and clarify his preferences.[21]

Similarly representing the states as vectors of parameters, structuring the representation of beliefs as hierarchical probability distributions or mixtures of models, and assimilating new information according to Bayes' Theorem counter all the biases inherent in unguided human judgement.[33,34] To use the Bayesian terminology, it guides the decision maker towards coherence.

Perhaps the greatest advantage of the Bayesian methodology is that it separates the issues of uncertainty and conflicting objectives within a problem. Without the structure of the decision analysis, the decision maker has to face up to both problems together and risk each confusing the other.

Throughout the above we have been arguing that Bayesian decision analysis helps the decision maker reach a decision without falling into all the traps of poor intuitive judgement. The structure of the analysis has a "debiasing effect".[41,42] In fact, a decision analysis probably guides a decision more through two other effects. First, the analysis should bring understanding.[21,23] Through sensitivity analyses (about which we have said alarmingly little), through exploring the independence/dependence structure inherent in the decision makers' beliefs and preference, and through simple devices sich as decision trees, the decision maker explores his problem and his own views of that problem until he understands all the ramifications of the choice that faces him. Second, we have referred to a decision maker. Usually, a group, not a single person, is responsible for a decision. Communication between members of the group is important, if an agreed decision is to be made. Without the structure of decision analysis, there is a danger of ineffective communication. Group members typically argue for alternatives, rather than objectives. Their debate often clouds rather than consolidates areas of agreement. If the discussion is organised upon the framework of a Bayesian decision analysis then areas of agreement and disagreement can be identified. Moreover, careful use of sensitivity analysis can discover the importance of disagreements: for instance, it is possible that two people have quite different subjective probabilities yet agree upon the choice of action.

REFERENCES

[1] L. J. Savage (1954) *The Foundations of Statistics*. Wiley, New York.
[2] J. Von Neumann and O. Morgenstern (1947) *Theory of Games and Economic Behaviour*. 2nd Edition. Princeton University Press.
[3] A. Wald (1950) *Statistical Decision Functions*. Wiley, New York.
[4] B. De Finetti (1937) "Foresight: its logical laws, its subjective sources" translated and reprinted in *Studies in Subjective Probability* (Eds. H. E. Kyburg and H. E. Smokler) 93–158. Wiley, New York, 1964.
[5] F. P. Ramsay (1926) "Truth and probability" in *The Foundations of Mathematics and other Logical Essays*. Kegan Paul, London, 1931.
[6] D. V. Lindley (1971) *Making Decisions*. Wiley, Chichester.
[7] H. Raiffa (1968) *Decision Analysis: Introductory Lectures on Choices under Uncertainty*. Addison-Wesley, Reading, Mass.
[8] M. Allais and O. Hagen (1979) *Expected Utility Hypotheses and the Allais Paradox*. D. Reidel Pub. Co. Dordrecht.
[9] P. H. Farquhar (1983) Utility assessment methods. In P. Hansen (Ed.) *Essays and Surveys on Multiple Criteria Decision Making*. Lecture Notes in Economics and Mathematical Systems No. 209. Springer-Verlag, Berlin.

[10] R. L. Keeney and H. Raiffa (1976) *Decisions with Multiple Objectives*. Wiley, New York.

[11] T. S. Wallsten and D. V. Budescu (1983) "Encoding subjective probabilities: a psychological and psychometric review". *Management Science* **29**, 151–174.

[12] P. G. Moore and H. Thomas (1976) *The Anatomy of Decisions*. Penguin, Harmondsworth.

[13] J. T. Buchanan (1982) *Discrete and Dynamic Decision Analysis*. Wiley, Chichester.

[14] R. L. Keeney (1981) Measurement scales for quantifying attributes. *Behavioural Science* **26**, 29–36.

[15] P. Byer and R. De Neufville (1978) "Choosing the dimensions and uncertainties of an evaluation" in *Formal Methods in Policy Analysis* (D. W. Bunn and H. Thomas, Eds). Birkhauser Verlag, Basel.

[16] R. L. Keeney (1981) "Analysis of preference dependencies among objectives". *Operations Research* **29**, 1105–1120.

[17] T. L. Saaty (1980) The Analytical Hierarchy Process.

[18] A. Goicoechea, D. R. Hansen and L. Duckstein (1982) *Multi-objective Decision Analysis with Engineering and Business Applications*. Wiley, New York.

[19] S. French (1985) *Decision Theory: An Introduction to the Mathematics of Rationality*. Ellis Horwood, Chichester.

[20] P. H. Farquhar (1980) "Advances in multi-attribute utility theory". *Theory and Decision* **12**, 381–394.

[21] S. French (1983) "A survey and interpretation of multi-attribute utility theory" in *Multi-Objective Decision Making* (S. French, R. Hartley, L. C. Thomas and D. J. White, Eds.). Academic Press, London, 263–277.

[22] P. H. Farquhar (1982) "Multivalent preference structures". *Mathematical Social Sciences* **1**, 397–408.

[23] P. C. Fishburn and R. L. Keeney (1975) "Generalised utility independence and some implications". *Operations Research* **23**, 928–940.

[24] R. F. Meyer (1977) "State dependent time preference" in *Conflicting Objectives in Decisions* (D. E. Bell, R. L. Keeney and H. Raiffa, Eds.). Wiley, New York, 232–244.

[25] R. F. Meyer and J. W. Pratt (1968) "The consistent assessment and fairing of preference functions". *I.E.E.E. Trans. on Systems Science and Cybernetics* SSC-4, 270–278.

[26] A. P. David (1979) "Conditional independence in statistical theory (with discussion)". *J. Royal Statistical Society* **B41**, 1–31.

[27] D. V. Lindley and A. F. M. Smith (1972) "Bayes estimates for the linear model (with discussion)". *J. Royal Statistical Society* **B34**, 1–41.

[28] I. J. Good (1980) "Some history of the hierarchical Bayesian methodology" in *Bayesian Statistics*. (J. M. Bernardo, M. H. De Groot, D. V. Lindley and A. F. M. Smith, Eds.). University of Valencia Press, 489–504.

[29] P. J. Harrison and C. F. Stevens (1976) "Bayesian forecasting (with discussion)". *J. Royal Statistical Society* **B38**, 205–247.

[30] A. P. David (1982) "Intersubjective statistical models" in *Exchangeability in Probability and Statistics* (G. Koch and F. Spizzichino, Eds.). North-Holland, Amsterdam.

[31] P. Diaconis and D. Freedman (1984) "Partial exchangeability and sufficiency". *Sankhya* (to appear).

[32] A. F. M. Smith (1984) "Bayesian statistics: a provocative term for disciplined common sense?" Paper presented at the Royal Statistical Society's 150th Anniversary Conference, April 1984.

[33] D. Kahneman, P. Slovic and A. Tversky, Eds. (1982) *Judgment under Uncertainty: Heuristics and Biases*. Cambridge University Press.

[34] R. M. Hogarth (1980) *Judgment and Choice.* Wiley, New York.
[35] A. Tversky and D. Kahneman (1981) "The framing of decisions and the psychology of choice". *Science* **211**, 453–458.
[36] D. Kahneman and A. Tversky (1979) "Prospect theory: an analysis of decisions under risk". *Econometrica* **47**, 263–291.
[37] C. R. Schwenk and H. Thomas (1983) "Effects of conflicting analyses on managerial decision making: a laboratory experiment". *Decision Sciences* **14**, 467–482.
[38] A. Tversky (1969) "Intransitivity of preference". *Psychological Review* **76**, 31–48.
[39] G. E. Wells (1982) "The use of decision analysis in Imperial Group". *J. Operational Research Society* **33**, 313–318.
[40] P. Slovic and S. Lichlenstein (1971) "Comparison of Bayesian and regression approaches to the study of information processing in judgment". *Organisational Behaviour and Human Performance* **6**, 649–744.
[41] B. Fischoff (1982) "Debiasing" in Reference 33, 422–444.
[42] D. Berkeley and P. Humphreys (1982) Structuring decision problems and the 'bias heuristic'. *Acta Psychologica* **50**, 201–252.
[43] L. D. Phillips (1984) "A theory of requisite decision modelling" submitted to *Acta Psychologica.*

4.4   GROUP DECISION MAKING AND DECISION CONFERENCING

I hope that I have justified my claim that decision analysis is a very subtle, interactive tool for exploring decision problems. It provides a framework for thinking about choices: one that catalyses the evolution of a decision maker's perception, beliefs and preferences. It can be a very creative tool in that, as a decision maker comes to understand the choice before him, he may discover new options and construct new strategies. In the MEM case study (section 3.6), the Managing Director's realization of the importance of a clear introduction and the value of spending a significant sum – in principle, up to £11·6 million – is a small illustration of the way in which this can happen.

I have also suggested that decision analysis can facilitate group communication, concentrating the members' efforts on relevant issues and helping them to a shared understanding of the opportunities before them. The MEM case study shows this in a number of respects. The likelihood of a ban, seemingly very important before the analysis, was appreciated by all to be irrelevant to the final choice: thus sterile discussion of this issue was avoided. (Note, however, that this was only true because the decision tree format had led the Managing Director to think creatively about all his strategies in the event of a ban.) In the sensitivity analysis one set of 'extreme' pessimistic probabilities and criterion weights which led to a revised choice was discovered. But no individual board member could accept all of these together, although most could accept some of the judgements inherent in these values. Thus each was able to see a decision analysis tailored to their individual views, but all were able to see that, despite their differences, their conclusions were the same. Potentially heated, unnecessary debate was avoided.

Phillips (1984) and Hall (1986) present further examples of the use of decision analysis to support groups of decision makers. They also identify two factors crucial to the success of group decision support, which we have yet to consider: the need to understand group processes and the power of modern microcomputers with their associated information technology.

If decision analysis is to support groups in the ways indicated above, then ideally it should be carried out within the group. The model should be developed and analysed in the group environment. The members need to see, understand and be able to contribute in their own language throughout the modelling process. The power of modern portable microcomputers with their graphical capabilities combined with overhead or video projection means that decision analyses can be carried out anywhere on the spot. Thus, increasingly decision analysts – or **facilitators**, as they are becoming known – are finding themselves in boardrooms working with management teams to solve problems in a matter of hours or days.

This close interaction between the facilitator and the decision-making team means that he has to be skilled in helping groups work effectively, as well as in all the decision-analytic tools that he will need. Group activity can be immensely creative and productive; alas, it can also be stultifying. Creating the right environment and mood for the former may test the skills of the best facilitator, but, unless he is successful, the power of the analysis may be greatly devalued.

Decision conferencing is one format which seems to be successful in achieving the right environment for effective group decision making. It was developed in the United States by Cam Peterson, and has been introduced into the UK and Europe by, among others, Larry Phillips and Peter Hall. The process is described in detail in Hall (1986).

Briefly: a decision conference is a two-day event in which all the owners of a problem gather together with three support staff:

a *facilitator* – the lead analyst, someone who is experienced in group processes and decision theory who helps the group to focus on their task, identify the issues, model the problem and interpret the results

an *analyst* – a support analyst, who attends to the computer modelling

a *recorder* – who uses a projected word processor to help the group capture their description of the issues before them, records the rationale behind their judgements and provides a record of the decision conference for the participants to take away with them.

Strictly, a decision conference need be long enough to solve the problem before it: no longer and no shorter. Thus it need not be two days long. However, an overnight break for informal discussion and individual reflection has invariably been found invaluable. Time constraints usually prohibit working consecutively with a senior management team for longer than two days. So, in practice, decision conferences have lasted two days.

These remarks, although clearly insufficient to describe decision conferencing fully, should indicate that there is much more to successful decision analysis than merely understanding and applying decision theory. Moreover, it should be admitted that the format of decision conferences is evolving with experience; to date, the majority have concerned situations in which uncertainty was not the central issue. Thus they have used models appropriate to decision making under certainty. Although she does not use the term, the models and style of working described by Belton in section 3.7 are typical of decision conferences as they are currently practised.

We have said little about the software available for decision analysis, and a book such as this is probably not the best place to do so. Humphreys and Wisudha (1987) have recently produced a wide-ranging survey of decision-analytic software. Bodily (1985) is a useful reference in that it is both an introductory textbook and a description of how to perform decision analyses with the IFPS financial and business modelling language.

4.5 SUBJECTIVE ASPECTS OF THE ART OF DECISION ANALYSIS

The paper by Thomas and Samson, reprinted below discusses further how decision analysis can help bring structure into unstructured problems. It also discusses many other issues that are of importance to a successful decision analysis.

Subjective Aspects of the Art of Decision Analysis: Exploring the Role of Decision Analysis in Decision Structuring, Decision Support and Policy Dialogue

HOWARD THOMAS and DANNY SAMSON
Department of Business Administration, University of Illinois at Urbana–Champaign

The paper argues that, until very recently, decision analysts have devoted relatively little attention to the processes of problem formulation and subjective judgement in handling ill-structured strategic decision problems. Therefore, following a brief review of existing varieties of decision analysis, a modified 'policy dialogue' model of decision analysis is presented which integrates decision analysis with decision aids and decision support technology. This model is developed using as an illustration strategic problems drawn from the insurance industry. The paper concludes with some suggestions for the successful application and implementation of decision analysis.

Key words: analysis, decision, information systems, insurance, modelling, planning, risk

INTRODUCTION

Formal approaches to organizational decision-making have been rarely applied, apart from a restricted set of techniques applied to specific operational problems. Indeed, concern has been expressed in the management science literature regarding both the breadth of applications and the rate of acceptance of consequent recommendations (Schultz and Slevin,[1] Ackoff[2,3] and Eilon[4]). In the decision analysis context, Kunreuther and Schoemaker[5] argue that when decision theory analysis is viewed as a multi-stage model for rational choice among alternative options, its impact on organizational theory and managerial behaviour tends to be less than might have been hoped for or expected (Behn and Vaupel,[6] Grayson,[7] Brown[8]). The limited attention given to the descriptive aspects of problem formulation (Hogarth[9]) and the inherently political nature of organizational decision-making has often been cited as the cause of the relatively limited adoption of decision analysis approaches.

However, numerous examples exist to demonstrate that decision analysis has been usefully and successfully applied to the analysis of such well-structured, well-specified situations as, for example, new product decisions, manufacturing investment, and oil and gas drilling decisions (Brown,[8] Brown et al.,[10] Grayson,[11] Moore et al.,[12] Kaufman and Thomas[13]). More recently. Keeney,[14] Keeney and Raiffa,[15] Kaufman and Thomas[13] and Ulvila and Brown[16] report an increase in the applications of decision analysis to complex, difficult, ill-structured problems and argue that decision analysis is especially valuable in such situations. Its extended use in both the corporate and the public policy areas (see, for example, Howard and Matheson[17]) suggests it may yet fulfill its potential as a useful decision aid for the formulation and analysis of complex problems.

It is argued that certain adaptations of the basic 'rational choice' decision analysis paradigm are required for it to be effectively applied to strategic decision and policy situations. In particular, the existence of structural uncertainty means that much attention must be focused on problem structuring and formulation. Therefore, decision analysis is presented here as a vehicle for generating dialogue about problem assumptions, formulation and available options, rather than as a means for the determination of an optimal strategy. This modified decision analysis approach is regarded as a support system for problem solving rather than as an optimal statistical technique.

Thus the paper is structured as follows. The modifications necessary to apply decision analysis to ill-structured problems (see McCaskey[18] and Mason and

Mitroff[19]) are outlined initially and illustrated using existing consultancy models of decision analysis. Particular attention is given to the role of decision analysis in policy dialogue. This is followed by some discussion of work undertaken in the insurance industry which illustrates the links between decision analysis and decision support and also highlights some implementation problems. Attention is then focused upon the need for analysts to develop clinical skills and strategies in order to increase the probability of acceptance and successful implementation of the ensuing policy recommendations. The paper concludes by summarizing the important features of the policy dialogue framework for decision analysis.

APPLYING DECISION ANALYSIS TO ILL-STRUCTURED PROBLEMS

The decision analysis approach (Raiffa,[20] Moore and Thomas,[21] Keeney[14]) is normally applied in terms of a series of distinct steps or stages (see Figure 1). These are:

(i) *Structuring the problem*: definition of the set of alternative strategies; the key uncertainties; the time horizon and the attributes or dimensions by which alternatives should be judged.

(ii) *Assessing consequences*: specification of impact or consequence measures for the decision alternatives.

(iii) *Assessing probabilities and preferences*: assessment (or definition) of probability measures for key uncertainties and utility measures to reflect preference for outcomes.

(iv) *Evaluating alternatives*: evaluation of alternatives in terms of a criterion for choice, such as the maximization of expected utility.

(v) *Sensitivity analysis* in relation to the optimal strategy, which may lead to further information gathering.

(vi) *Choice* of the most appropriate strategy in the light of the analysis and managerial judgement leading to implementation of the preferred strategy.

Since this basic paradigm was proposed, the experience gained by both consultants and academics has stimulated changes designed to make the decision analysis approach more flexible to the needs of managers. In many applications the attention has moved away from the 'purity' of the analysis and the search for an optimal solution. Instead, the focus is more frequently upon such factors as the 'mess' (Ackoff[22]), the complexity, and the bargaining, debate process which characterize so many ill-structured policy and strategy problems. Indeed, such consultancies as Woodward–Clyde in San Francisco, Decisions and Designs (D.D.I.) in Washington, Decision Science Consortium (D.S.C.) in Washington and Stanford Research Institute in Menlo Park have adapted their versions of decision analysis to the realities of the market-place and the increasingly illstructured problems which they seek to resolve.

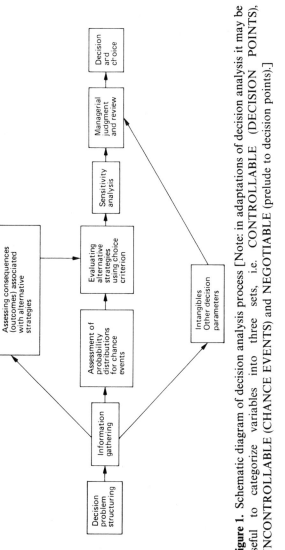

Figure 1. Schematic diagram of decision analysis process [Note: in adaptations of decision analysis it may be useful to categorize variables into three sets, i.e. CONTROLLABLE (DECISION POINTS), UNCONTROLLABLE (CHANCE EVENTS) and NEGOTIABLE (prelude to decision points).]

Step 1. Structure the problem

      Identify the alternatives

      Specify relevant impact groups

      Determine the objectives

      Define measures of effectiveness (attributes) for
      each objective

Step 2: Assess the possible consequences of the alternatives

      Quantify consequences in terms of attributes

      Assess judgments of experts

      Collect data and update estimates

      Quantify uncertainty, using probability

Step 3. Determine the preference (value) structure

      Determine the general form of the utility function to
      quantify the value structure

      Assess the single attribute utility functions

      Assess the value trade-offs to indicate relative
      importance of different objectives

      Verify the consistency of the value judgments

Step 4. Evaluate and compare the alternatives

      Integrate the previous information to evaluate
      alternatives

      Conduct a sensitivity analysis with respect to
      preferences and consequences

      Re-examine aspects found to be crucial
      to the decision

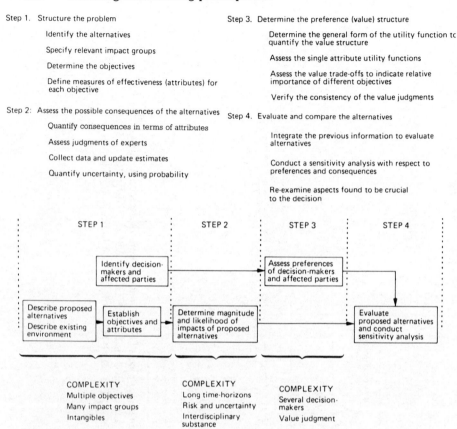

Figure 2. Steps in decision analysis. From R. L. KEENEY (1979) Decision analysis: how to cope with increasing complexity, *Mgmt Rev.* p. 26.

It is useful to examine how these consultancies have used the decision analysis approach and have developed distinct styles in relation to their differing areas of application. For example, Figure 2 shows the steps in decision analysis as conceived by the Decision Analysis Group at Woodward–Clyde Consultants, in which Ralph Keeney (now at the University of Southern California) and Craig Kirkwood (now at Arizona State University) were perhaps the most well-known principals. The group has worked most closely with problems in the environmental, regulatory, social and legal areas, such as the siting of energy facilities. Typically, these problems involve *high stakes*, have *complicated structures*, need *multiple viewpoints* for resolution (i.e. there is no single expert). In addition, the decision-makers are usually required to justify decisions to regulatory authorities, corporations and the public at large.

Figure 2 notes the complexities of such problems which require the adaptation of the basic, single decision-maker, Raiffa-type paradigm. As a result of law or regulation, they involve the consideration of multiple objectives and involve many impact groups. They have long time-horizons, are characterized by significant uncertainties and involve many decision-makers, who are forced to recognize the interdisciplinary substance of the decision situations. Using such approaches, Woodward–Clyde have generated considerable academic and practical research in the application of the technology of multi-attributed utility theory (MAUT) (Keeney and Raiffa[15]) to the class of decision problems involving the interface of the organization with legal, regulatory, social and economic environmental forces. Such cost–benefit type analyses require a keen awareness of strategic management of interorganizational forces and typically use MAUT as the input to a debate process concerning policy choice. These implementation concerns are familiar to generations of cost–benefit analysts working on applications in the area of welfare economics.

In contrast, the Decision Analysis Group at S.R.I., originally founded by Professor Ronald A. Howard and Dr James Matheson (now principals with the Strategic Decisions Group in California), has a much greater model-building emphasis than the other consulting groups. They structure their version of the decision analysis process in terms of the decision analysis cycle shown in Figure 3. The deterministic phase calls for problem formulation, structural modelling, the specification of value and time preferences and, particularly, extensive sensitivity analysis, which provides the link between the deterministic and probabilistic phases. The probabilistic phase introduces probability distributions for certain key numerical and structural factors and generates a probability distribution for a performance criterion such as net present value (N.P.V.), which displays the perceived risk of various alternative strategies. The determination of certainty equivalents for these distributions enables value judgements in relation to risk to be made. The informational phase stresses the economic value to be obtained from reducing the uncertainty characterized in the probabilistic phase. Additional information-gathering may thus be deemed uneconomic in terms of a cost–benefit trade-off between time and money (Howard[23]).

Thus S.R.I. sees decision analysis as an interactive process. In essence, the simplest initial analysis (i.e. deterministic) consistent with the structural model should be carried out. This is referred to as a pilot-level analysis. Guided by the results of this analysis, a more detailed prototype study (involving probabilistic analysis) is undertaken. If deeper analysis and system sensitivity analysis are required, then a final stage 'production' level of analysis can be generated. The economics of information gathering is seen to be controlled by analyses of the value of such information. The S.R.I. approach is perhaps more engineering- and systems-oriented than other approaches. Models are seen as providing a road-map for decision logic and allowing, through the process of decomposition, various information sources to be very specifically targeted. They thus involve a capital investment beyond the use for which they were originally constructed. They provide a basis for ongoing decision analyses and for a continuing client–consultant relationship.

Stanford Research Institute has worked extensively on both commercial and public sector analyses. These range from company-wide planning and strategy models to such public sector applications as decisions to 'seed' hurricanes and plan fire protection for the Santa Monica mountains in Los Angeles (Kaufman and Thomas,[13] Howard and Matheson[17]). Their work increasingly involves ill-structured problems, and their operational paradigm is to break the decision problem into its constituent parts, reassemble them in a step-by-step approach and improve the analysis in a cyclical way.

In common with many of the newer decision analysis consultancies such as the Decision Science Consortium, D.D.I. perhaps represents a more process-oriented decision analysis technology relative to the more focused modelling approach of S.R.I. Therefore, D.D.I. focus much more attention on the process of generating alternatives and of helping decision-makers to structure alternatives. They often use extremely simple linear additive MAUT approaches suggested by Edwards[24] to identify alternatives, sometimes using a user-friendly interactive MAUT package (Humphreys and Wisudha[25]) to clarify the hierarchy of value attributes relevant to the problems. D.D.I.'s use of linear models can be seen as a decision simplification mechanism to enable decision-makers to develop better alternatives. Successful passes of this process are used to develop a better understanding of assumptions and a more realistic set of alternatives.

Essentially, D.D.I.'s view of the world argues that a decision analyst does not have to structure the entire problem. Decision analysis is seen as most useful in describing and debating possible implications of some aspects of the problem rather than optimizing with respect to all of them. Rex Brown (now at the Decision Science Consortium) and Cam Peterson cite a D.D.I. analyst who gives an example

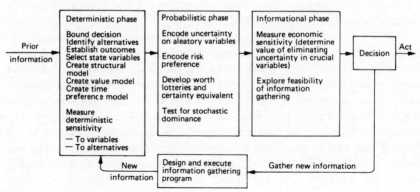

Figure 3. From Kaufman and Thomas,[13] p. 119.

concerning the use of decision analysis in comparing alternative disarmament strategies. The analyst focused attention solely upon the prediction of how long it would take NATO to mobilize given a Warsaw Pact attack. Decision analysis was used as a descriptive device. That is, NATO was assumed to make the mobilization decision as a rational unitary actor, and then the assumptions were relaxed and debated so that decision-makers could accommodate the analysis to the actual operations of complex bureaucratic processes. Thus, D.D.I. used a mixed-scanning multiple viewpoint approach to top-level decision-making suggested by such writers as Allison[26] and Etzioni.[27]

Much of D.D.I.'s applied experience has been in the public sector and, especially, in military and defence applications. The complexity of such public-sector problems has probably influenced changes in their analytic style, involving the use of decision analysis as a decision-aiding technology and a decision support system rather than simply a solution technology.

Thus it is argued here that these consultancies have developed and refined decision analysis. There is less concern about methodological issues such as probability and utility assessment. More attention is now given to aiding the decision-maker in problem formulation, screening of alternative options and in promoting effective dialogue about problem characteristics and policy issues. In other words, the important principle in modifying decision analysis should be that formulation and evaluation of ill-structured problems requires a creative mix of analytic inputs and continual debate. It must be recognized that it is almost impossible to undertake anything other than an exploratory and preliminary analysis at the first attempt. If so, this 'first pass' analysis should be documented and subjected to critical comment and review by the policy-making group. In the course of this process, debate about the problem will become more focused around questioning of assumptions, generation of further alternatives and anticipation of future contingencies.

As Keeney and Raiffa[28] say:

> Simply stated, the major role of formal analysis is "to promote good decision-making". Formal analysis is meant to serve as an aid to the decision-maker and not as a substitute for him....
>
> ... As a process, it is intended to force hard thinking about the problem area: generation of alternatives, anticipation of future contingencies, examination of dynamic secondary effects, and so forth. Furthermore, a good analysis should illuminate controversy – to find out where basic differences exist, in values and uncertainties, to facilitate compromise, to increase the level of debate and undercut rhetoric – in short, "to promote good decision-making".

This modified decision analysis is an approach and a broad investigative research strategy rather than a technique, and is not necessarily performed in a series of sequential steps. Some steps may be excluded or handled in an informal manner. The order of the steps may be varied and, indeed, the relevance of the objective structure, problem assumptions and the importance of excluded factors may be continually reassessed. In particular, the philosophy of the modified decision analysis approach strongly emphasizes the point that the identification of new options is even more important and necessary than anchoring firmly on analysis and evaluation as goals of the analysis.

Clearly, this modified decision analysis approach can still incorporate the techniques outlined in Figure 1. Indeed, when dealing with well-structured problems (such as oil and gas exploration), the traditional and modified paradigms are identical. The value of the new paradigm is embedded in its flexibility, which is needed to deal with increasingly complex and unstructured policy and strategic management issues. It requires deliberate and disciplined use, but yields greatly enhanced understanding of the nature of the problem and the available options.

The goal of this modified approach should be judged in terms of its contribution to organizational processes rather than specifically recommending an action and getting it adopted. Very often the understanding derived from the process of structuring the problem and the information related to outcomes and actions may significantly influence the quality of the decision process.

DECISION ANALYSIS AS AN AID FOR POLICY DIALOGUE

The previous sections of this paper argue that analysts have increasingly sought to build flexibility and adaptability into the analytic process. In nearly all applications, an initial, very preliminary model of the decision situation is developed which gives decision-maker(s) an opportunity to explore and understand the problem situation more clearly. For example, it may often be preferable for decision-makers to examine the outcomes of alternative policies in terms of a time-stream of indicator variables, rather than in terms of a single criterion such as expected utility, whether or not that utility function is expressed in multi-attributed form. In this manner, policies can be discussed and debated even more thoroughly by managers to see whether they fit in terms of a broad spectrum of indicators. This is particularly important when excluded factors and problem assumptions are re-examined. Discussion typically leads to the advocacy of different policies and views of the world. Consensus about problem formulation can only be achieved through an inquiry system which encourages strategic dialogue about the consequences of alternative assumptions, problem formulations and scenarios.

In practice, the application of decision analysis to ill-structured strategic problems sometimes involves an effective policy dialogue about alternative options. Examples are available in Bunn and Thomas,[29] Thomas,[30] Holling,[31] Bell,[32] Meyer,[33] Keeney and Raiffa,[15] Keeney,[14] Ulvila and Brown,[16] Hertz and Thomas[34,35] and Howard and Matheson.[17]

Perhaps the clearest meassage from these more recent applications of decision analysis is that there is no unique way to ensure that the problem is adequately structured and the set of probabilities are well assessed. Decision-makers and organizations vary greatly. Analysts need to be flexible and creative in applying decision analysis. Therefore, the approach adopted must recognize the character-istics of the organization, the ambiguity inherent in the decision problem and the training, experience and personalities of the key decision-makers.

However, it seems sensible to ask whether confusion or clarity is generated by advocating the use of combinations of planning and analytic approaches to develop alternative and often conflicting problem viewpoints for policy dialogue. In a laboratory experiment involving business executives, Schwenk and Thomas[36] provide evidence demonstrating that the presentation of conflicting analyses is more effective than a single analysis in improving decision-making performance.

Finally, the advent of available computer and communication technology has enabled sophisticated decision analysis models to be provided in a 'video-conferencing' mode using the concept of managerial decision support. The development of user-friendly software for assessing probabilities and utilities and observing the potential impacts of alternative strategic options in graphical terms has also strengthened the viability of the concept of decision analysis as a strategic inquiry system (Churchman[37]). Some of the available decision support systems for automating the process of decision analysis are discussed in the next section. Probability assessment packages such as Schlaifer's MANECON package[38] and S.R.I.'s[39] Automated Aids for Decision Analysis are reviewed briefly, followed by an example of a strategic decision support system drawn from the authors' recent research in insurance (Samson and Thomas[40]).

DECISION SUPPORT SYSTEMS FOR DECISION ANALYSIS: A BRIEF REVIEW AND AN EXAMPLE

The MANECON programs (Schlaifer[38]) provide a suite of programs which help decision-makers to ensure consistency of subjective assessments of probability and utility. Although not immediately user-friendly, they have been adapted in doctoral theses by researchers such as Peter Burville (London) and Michael Middleton (Stanford) to provide on-line probability assessment procedures in decision-making contexts.

The research of Stanford Research Institute's Decision Group[39] consciously attempts to widen the focus of decision analysis away from the traditionally narrow orientation involving the stress upon such factors as assessment quality and consistency. They emphasize the logical and analytical steps necessary for the analysis of a wide variety of decisions and link decisions with the new technologies of decision support systems. This decision support focus, which is echoed also in Schwenk and Thomas's[41] paper, has produced:

(1) a characterization of the different kinds of decision situations that arise in

practice and an exploration of the implications of these characteristics for automated decision aids;

(2) a description of the types of decision models available for analysing a variety of decision situations;

(3) a description of the process of constructing decision models; and

(4) an identification of several easily understood modelling concepts that provide a basis for designing and constructing a pilot-level system of automated decision aids.

Other researchers have also focused on the role of decision analysis approaches as aids in decision-making. For example, Humphreys[42] has developed the MAUD system for automating multi-attributed utility assessment in decision-making, and Herbert Moskowitz of Purdue has produced a user-friendly program for assessment of probabilistic scenarios in complex decision-making situations. Perhaps one of the main problems associated with forecasting 'fuzzy' futures is the need to assess adequately key scenarios and the assumptions which underlie the construction of such scenarios. Hertz and Thomas[34] also report that the probability tree, or 'fault' tree, as it is commonly referred to by engineers, is a very useful aid for structuring the thinking process, provided that the decision-maker is encouraged to think about the range of possible outcomes. A number of corporate-planning groups which encourage scenario construction for 'futures' have identified 'anchoring' bias around central or 'status quo' values, and have modified assessment procedures to avoid asking for the 'most likely' scenario.

Many other suggestions for reinforcing decision analysis's role in decision support could be presented here (see, for example, Slovic[43]). Instead, a practical example of the comprehensive development of decision support involving decision analysis is discussed below. This example illustrates practical problems of assessment and implementation in the context of the insurance industry.

DECISION SUPPORT AND INSURANCE

The decision support system discussed here describes strategic planning problems commonly faced by many insurance companies. In addressing these problems, elements of subjectivity enter the problem and model-structuring phases as well as the numerical assessment phase. Indeed the process of generating alternative strategic plans requires experience and creativity, and hence is more effectively performed by managers than by their computers. However, when managers are efficiently supported by computers, their problem-solving ability can be increased. The effectiveness of their subjective judgements can be increased as a result of the high quality of analytic information supplied by the computer. The decision support system described below is one which provides building blocks (i.e. modules) from which the manager constructs models of future strategic scenarios. The computer then evaluates each of these scenarios and, based on these evaluations, the manager and computer can interact in an iterative manner in an attempt to create improved strategies (as measured by an appropriate utility function).

STRATEGIC PROBLEMS IN INSURANCE

Some important strategic questions which have to be considered in insurance organizations include the following:

(1) What portfolio of types of insurance (lines of business) should the company underwrite and what operating strategies should be adopted for each line of business?
(2) How should premium income and other assets (known as reserves) be invested given uncertainties in stock, bond and option markets and other investment opportunities?
(3) What reinsurance arrangements should be made to spread corporate risk?

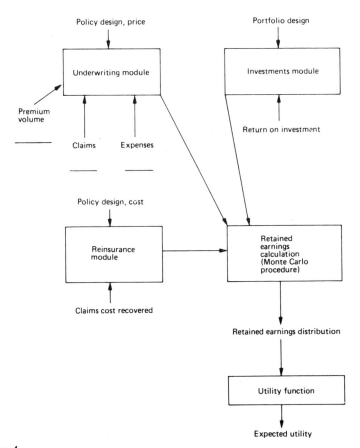

Figure 4.

Individually these decisions are quite complex, for they involve interrelated variables, as well as uncertainty about problem structure and the values of key parameters. Clearly, analysis of these problems in isolation constitutes a form of so-called suboptimization, and the solution thus generated may not be sensible when the insurance organization is viewed as a corporate portfolio of activities.

However, recent computer developments have allowed the corporate portfolio problem to be more simply analysed. For example, a modular decision support system (such as that shown in Figure 4) can provide the user with flexibility in structuring as well as enabling the inclusion of relationships between variables (where such relationships may be deterministic or probabilistic) and relevant problem constraints. Such a system allows top management to expand their role in corporate planning, and particularly in the stages of strategy generation and evaluation.

The data requirements of the system are shown in Table 1. For uncertain quantities, probabilities can be assessed as discrete values (essentially in histogram form), or distributions can be called upon from a bank stored by the system, in which case parameters must be specified (see Spetzler and Stael Von Holstein[44] and Moore and Thomas[45] for assessment methods).

The D.S.S. can be used as a model for the examination of retained earnings as a function of various sets of decision variables. The decision analysis approach also allows for the determination and use of a preference (utility) function for the user as part of the system. The retained earnings expression is described in an expanded form in the Appendix. Figure 4 also provides an overview of the D.S.S.

The underwriting module requires the input of policy design and price variables and estimates in the form of probability distributions for premium volume, claims

Table 1. Variables included in the decision analysis

Profit centre	Major decision variables	Major sources of uncertainty
Underwriting activities	Policy design variables Premium Types of insurance offered	Claims Expenses
Investment activities	Total funds to be invested Proportions of mix in various types of instruments Specific fund flows	Returns on investment stock prices dividends variable rate bond yields real estate values, etc.
Reinsurance indemnity activities	Types of reinsurance Reinsurance policy basis Extent of reinsurance Policy design variables	Claims on policies Cost of reinsurance Ability to recover on claims

and expenses. These inputs are made separately for each underwriting line or type of insurance.

The investments module requires input data involving investment opportunities and probability distributions about likely returns. From this data, an aggregate R.O.I. distribution is determined.

The reinsurance module requires inputs of reinsurance types, extents and costs (i.e. reinsurance premiums). Reinsurance can be of proportional or non-proportional form or a combination of both, and can be taken out either on a line-by-line basis or on an aggregate basis. All of these options can be evaluated using the D.S.S.

The output resulting from the three basic modules (underwriting, investments and reinsurance) is then used as an input in order to determine a probability distribution for retained earnings in the retained earnings module of the D.S.S. It should be noted that simulation approaches are typically used to generate the distribution for retained earnings because exact analytical solutions are only possible with particular forms of input distributions. Figure 5 shows a typical set of probability distributions (of retained earnings) obtained for three alternative strategies. The value of the overall strategic plan could be addressed in a number of ways. For example, the system can be used to develop managerial knowledge and

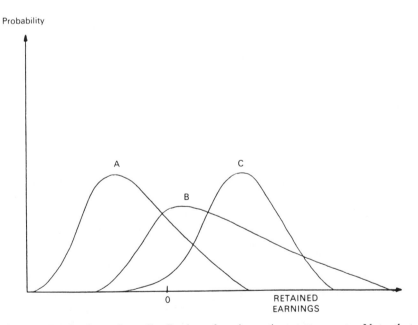

Figure 5. Retained earnings distributions for alternative strategy sets. Note that strategy A is stochastically dominated. In cases where stochastic dominance cannot be used to eliminate all but one strategy, expected utility rules can be used.

understanding about the likely results of underwriting, investment and reinsurance activities and their effects on performance measures such as retained earnings. Evaluation of strategies can also be accomplished through interpretation and debate about the expected utility of alternative strategic plans derived from the D.S.S.'s utility functions module. Alternatively, the principles of stochastic dominance may be applied to the retained earnings distributions to determine a feasible set of strategies.

CONSTRAINTS AND CONDITIONS ON PARAMETER VALUES

The overall aim of the system is to allow the user to determine values for those strategic variables important for maximizing the expected utility of retained earnings, subject to certain constraints and conditions. These constraints need not be explicitly included in the program, i.e. the user can check to see that they are satisfied for any set of variables prior to using the D.S.S. Examples of constraints may be those imposed by government regulators regarding minimum surpluses or premium/surplus ratios. Alternatively, they may involve internally imposed constraints on the composition of an investment portfolio, as, for example, bounds on the proportion of assets invested in bonds. However, the implications of these constraints can be included in the D.S.S. and examined comprehensively within the system for each set of input variables.

Many correlations exist between strategic variables which are not binding constraints, and these can also be either included or excluded from the D.S.S. An example is the relationship between claims adjustment expenses and reinsurance premiums. In general, increasing efforts (and hence expenses incurred) in claims adjustment would tend to lower claims and thus should lower reinsurance premiums. The relationship is a complex function of claims adjustment effectiveness, claim frequency and severity and reinsurance policy variables. The D.S.S. can be adapted to include such relationships and dependencies, or alternatively they can be accounted for externally to the system. A compromise option is to program the D.S.S. to monitor such relationships without knowing their explicit functional form. When a change is made to a variable of interest, appropriate correlations can be calculated for the user, thus allowing appropriate changes to be made to related variables. Thus, the system can be made more 'artificially intelligent', so that, for example, it is capable of suggesting reasonable responses in decision variable levels to changes in exogeneous conditions (such as trends in various insurance markets).

This type of D.S.S. may, therefore, be applied to any one of a large number of situations which are classified here as global (corporate-level) strategic planning, or business-unit-level planning. In both types of applications, many subjective judgements must be made about problem structure and the design of strategic actions as well as about the values of relevant variables and their associated probabilities. The user builds the structure and can perform 'structural sensitivity analysis'. In complex problems such as this one, there is not an objectively known correct structure, but rather a number of possible modelling structures and a number of alternatives to be evaluated. The model structuring and alternative generation processes are important subjective phases in decision analysis.

In a global strategic planning process, the user can examine the effects of alternative business strategies in combination with various assumptions regarding environmental variables such as market conditions, reinsurance prices and regulatory constraints. In this manner, the D.S.S. acts as both a structuring and an aggregative evaluation framework since it provides the user with information on the combined effect of many interconnected actions.

In many companies changes are often considered for one business line which may have secondary effects on other strategic activities. For example, a general insurer may be considering the introduction of a new policy type and a new product line. The D.S.S. can be used to examine the effects of different operating strategies by inputting various sets of prices and policy designs (along with accompanying probabilistic estimates for premium volume, claims and expenses). Secondary effects on other variables, such as the effect on investment funds of the new premium income, can also be assessed. The strengths of the D.S.S. in this case are in its ease of use and its ability to relate all policy changes to the aggregate financial performance of the firm (i.e. retained earnings). Since the system is based on the decision analysis paradigm, uncertainty is accounted for as well as the organization's attitude towards risk.

Recent technological advances have made computers available to many managers, and the processes of structuring and solving complex, messy problems can be aided and supported by the power of computers. The risk analysis approach and complex multivariable sensitivity analyses no longer pose computational difficulties even for complex problems. As a result, fewer simplifying assumptions are necessary in modelling processes, so that larger and more realistic models are being made available to the manager on his desk-top computer.

It is clear, however, that the increasing managerial use of sophisticated decision analytic models for complex problems places an increased burden upon managers properly to structure problems and assess key uncertainties. Our experience with decision support systems and decision-aiding processes in general has led to the conclusion that flexibility in problem definition and structuring is at least as important an issue as the design of appropriate subjective probability assessment procedures. Further, analysis takes place in a complex organizational decision-making process and requires analysts to develop clinical strategies to handle clients and facilitate problem definition and structuring.

DEVELOPING CLINICAL SKILLS AND STRATEGIES

Modifying the decision analysis paradigm requires consequent changes and improvements in the conduct of the implementation process. Based on experience in applying the decision analysis as policy dialogue approach, some implementation guidelines are suggested below. By way of introduction, Fischoff[46] considers the problems that may arise in decision analysis as a result of low awareness of clinical issues, and defines the skills relevant to psychotherapists as follows: "...they must instill confidence in clients, choose the appropriate questioning procedures to elicit sensitive information, handle crises, understand what is not being said, avoid imposing their own values and perceptions, and cooperate in creating solutions". The implication is that the skills required by decision analysts are similar.

The implementation problems are briefly discussed in the following paragraphs. A much more detailed discussion of these issues can be found in Lock and Thomas.[47]

THE INITIAL CONTRACT

During the initial stages of the decision analysis process, the analyst must explain the requirements of the process in terms of information requirements and the preferred degree of access to organizational decision-makers. In some more politicized organizations, sponsoring coalitions or individuals may attempt to control access to the analyst or the analyst's access to other participants. This should be recognized beforehand in developing the initial structuring of the problem and deciding which groups' views are essential in devising acceptable strategies and representing preferences. Interested groups might include a wide range of stakeholders, including owners (the state, shareholders, community, etc.), employees, consumers, managers and society. Yet, the incorporation of the views of other groups can lead to a decision analysis model rather different from that anticipated, or welcomed, by the sponsor (Kunreuther[48]).

The role adopted by the analyst

The analyst may interface with clients in a spectrum of possible roles, ranging from the 'expert' at one extreme to the 'trainer' at the other extreme. In the expert role, the structuring and analysis is largely performed by the analyst, without significant organizational involvement. In contrast, in the trainer role, the analyst aims to teach organizational decision-makers to structure analyses and evaluate alternative strategies on their own, and thus develop the ability to use the techniques in the absence of the analyst.

It should be noted that involvement and resulting commitment on the part of decision-makers increases as the analyst moves from the 'expert' to the 'trainer' role. The tendency for the conventional decision analysis paradigm to follow the first path partially explains the resultant low commitment to conclusions and recommendations in a number of case studies reported in the literature.

Diagnosing, exploring and structuring the problem

One view of the decision analyst is that of a passive encoder of client-provided information. However, this view assumes essentially that the organizational decision-makers have a fully developed understanding and representation of the decision problem. In complicated applications, problem formulation is frequently the most time-consuming phase (see, for example, Bunn and Thomas[29]). Despite the apparent critical importance of problem structuring and formulation in the strategic decision process literature (Mintzberg et al.,[49] Lyles and Mitroff[50]), mainstream decision analysis texts have tended to bypass it, suggesting that the process is more art than science. Others have also argued that problem structuring is learned by experience (Moore and Thomas,[21] Brown et al.[10]).

Recent studies in personal decision-making give much greater emphasis to the formulation process and state that much of the value of decision analysis seems to come from the structuring phase (Jungermann,[51] Humphreys[42]) when subjects' representations of the situations and problems are developed. Several specific areas in which probing by the analyst is valuable have thus been identified. One area is the elicitation of the range of goals and decision criteria. These in turn assist the definition of the range of alternative actions (Jungermann and von Vlardt[52]). The second area is exploring how actions are linked to outcomes. As well as specific questions of how different situation aspects are affected by particular actions, it is also necessary to identify who will be affected by a particular decision and their likely response.

The aim of the structuring phase is to generate an acceptable decision analysis model (Phillips[53]), which captures concisely problem elements and provides a problem description that can be discussed with the decision-makers to aid problem understanding. Subjectivity and creativity is required in model design so that only diagnostic events and critical trade-offs are retained. The remaining information gathered may be used in later sensitivity analyses.

From the above it may be seen that the process is a cyclical one, in which the technology is a structuring aid in itself (Thomas[30,54]). This process of formulating ill-structured strategic problems may also require specialized aids. A number of aids have been proposed to assist this process: for example, the concepts of creativity stimulants (Prince[55]); devil's advocate (D.A.) (Schwenk and Cosier[56]); dialectical inquiry (D.I.) and strategic assumptions analysis (Mason and Mitroff[19]); and finally, Delphi decision analysis (Wedley et al.[57]). Schwenk and Thomas[41] provide an integrative model incorporating decision aids into the decision analysis process (see Figure 6). They present a process by which a range of alternative scanning models (devil's advocate, dialectical inquiry, creativity stimulants and decision analysis as policy dialogue) can be used to develop a sound decision problem formulation. The process is presented as a cyclic search process in which decision-makers are encouraged to cycle back through previous stages of analysis. For example, either a structured debate (D.A. or D.I.) (cycle 2) and/or creativity stimulants (cycle 1) may be needed to reformulate a problem following a dialogue (cycle 3) about the initial decision analysis.

Approaches such as the devil's advocate and dialectical inquiry involve the introduction of conflict into the corporate problem formulation process. In

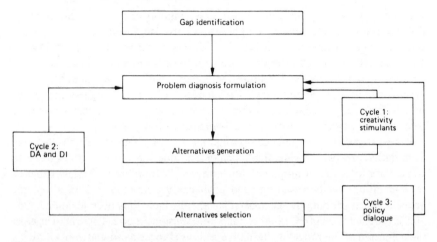

Figure 6. Appropriate uses of decision aids in the problem formulation decision–making process. Source: Schwenk and Thomas.[36]

particular, writers in the area (e.g. Mitroff and Emshoff[58]) suggest that three activities exist which can improve the quality of problem formulation in uncertain environments. The first is the generation of conflict between the decision-making group or within a decision-maker. The second is the identification of assumptions about the nature of the problem, and the impacts of the internal and external decision environments. The third is the challenging of assumptions.

If there is a potential drawback to the adoption of such processes, it seems that decision-makers may resist adoption because of the need continually to re-examine assumptions, even when a solution has been proposed. This continual re-examination imposes a time requirement which may not be feasible or acceptable and may open up areas which are regarded as particularly politically sensitive (Schwenk and Thomas[36]).

The analysis process: presenting solutions

Having arrived at an appropriate initial problem structure, assessment of the probabilities and preferences identified within the representation of the structured decision problem can proceed using appropriate encoding aids. Ultimately the 'first-pass' analysis has to be useful to decision-makers, and they have to feel confident about it. The presentation of a single best option does not always inspire this confidence. Strict optimization is less attractive than the ability to explore the problem through policy dialogue of several passes of the analytic model involving varying assumptions about problem elements. A major role for computer-based decision support models exists in the facilitation of manager-based sensitivity analyses and the ability to respond to 'what-if' questions.

Decision-makers acquire commitment to the solution by feeling both that they have some control over the policy recommendations and that they have contributed to its development.

IMPLEMENTATION: DEALING WITH CONFLICT

The main elements in devising an implementation strategy relate to identifying the key groups and individuals, how they can be induced to contemplate change and how they will respond to any particular proposals. The person with the perceived responsibility for the decision should be in charge of the implementation strategy rather than the decision analyst. Further, it is clear that decision-makers often dislike conflicts, particularly highly personal ones, and seek to avoid them. In many situations, changes and decisions are postponed until they are imposed by an external agency. By this time, the organization's survival may even be threatened.

The alternative is to consider to what extent it is possible to improve the client's or the client organization's ability to deal with conflict. Porter et al.,[59] Thomas[60] and MacCrimmon and Taylor[61] discuss various ways of resolving conflict. In cases where conflict is not directly resolvable, it may be feasible to assist people to handle overt conflict and to confront political issues openly through the use of conflict-based decision aids (Schwenk and Thomas[36]). This appears to be a development strategy that may require a longer time-horizon than is usually available in a decision analysis study.

SUMMARY AND CONCLUSIONS: THE POLICY DIALOGUE PARADIGM

The decision analysis dialogue paradigm (and the associated role of decision analysis as a decision support system) presented in this paper should be seen as a vehicle for a continuous policy dialogue involving analysis, assumptions and contingency planning. Examples using such dialogue processes have emerged from applications in the insurance industry, research and development, and in new product and diversification planning (Hertz and Thomas,[62] Lock and Thomas,[63] Samson and Thomas,[40,64] Thomas[65,66]).

In the context of developing and sustaining the policy dialogue, the following questions are relevant (see also Fischoff[46] and Slovic[43]).

(1) What are the assumption bases of the decision-makers and the analyst? If different, how do they know when they are arguing from different premises?
(2) On whose assumptions is the preliminary formulation based?
(3) What justification is there for the assumptions? Which internal sources might be used for data? Can the assumptions and data collected be checked and extended from external sources and environmental scanning? How accessible are the problem assumptions?

(4) Can one generate alternative problem representations? Which is most useful?
(5) What methods are available for assessments of probability and preferences? Which are most appropriate?
(6) How is analysis to be used in problem finding and solution? How should one perform the initial screening, and subsequent contingency and sensitivity analyses?
(7) To what extent are the variables interrelated? Can you perform some cross-impact analysis?
(8) Has appropriate consideration of political, legal and organizational factors been made?
(9) How should the analysis be evaluated?
(10) Is it possible to generate external criticism to improve the analysis?

It is argued that decision analysis as an aid for policy dialogue is a useful adjunct to other approaches for formulation and analysis of ill-structured problems. However, typically its value will be situation specific, and is likely to be both an adjunct to group discussion processes and an aid in clarifying policy evaluation and choice.

To quote Mason and Mitroff:[19]

> In our view, the task of policy, planning and strategy should not consist of attempting to demonstrate the superiority of one approach or framework for all situations but rather of showing their mutual dependency.... Whatever methods are used they should always aid in challenging strategic planning assumptions (p. 302).

The key theme in the dialogue approach is that, with messy problems, the initial problem formulation is very much a 'first pass'. By feeding back structured information from the 'first-pass' model, decision-makers will be able to improve both the range of options being considered and the representation of the relationship between options, critical exogenous variables and attribute outcomes. The emphasis is on a cyclical process where effort has to be made to avoid premature closure of any one phase. The skills involved in the representation process – modelling the option structure – are likely to come from a wide range of disciplines. Decision trees are but one way of approaching the modelling problem and can quickly be a cumbersome aid in modelling complex problems.

On the behavioural side, sensitivity to the organizational climate and the organizational consequences of any decision are likely to be crucial to both the likelihood of implementation and the success of such an implementation. The view of the role of the analyst as a change agent enables one to focus on the strategy that should be adopted in this role and the degree of client involvement that should be sought. This in its own way also tends to be an ideological issue reflecting the analyst's goals and their relationship with those of the client. The role of formal techniques in the policy and strategy framework is one of aiding organizational decision processes rather than supplying a single 'optimal' solution.

APPENDIX

Construction of the model begins with a set of accounting relations as follows:

$$\Delta A_D = \Delta C + \Delta R + \Delta RE. \tag{1}$$

The financial parameters of the firm can be represented as:

$$\Delta A_D = \Delta FA + \Delta DA + \Delta I + \Delta WC - \Delta LD, \tag{2}$$

where

A_D = total share capital and reserves at time n;

ΔA_D = change in A_D during period n to $n + 1$;

ΔC = change in paid-up capital during period;

ΔR = change in reserves (share premium, asset revaluation) during period (N.B. these are not claims reserves);

ΔRE = change in retained earnings during period;

ΔFA = change in fixed assets during period;

ΔDA = change in deferred assets during period;

ΔI = change in investments and loans during period;

ΔWC = change in working capital;

ΔLD = change in long-term debt.

The term which is most relevant to the decision process being studied is the retained earnings:

$$RE = (1 - T)[(1 + \lambda_n)P_n - X_n + R_n + PI_n - E_n - PR_n + CR_n]$$
$$+ DR_n - DP_n. \tag{3}$$

(Note that the expression for retained earnings, RE, is a highly simplified version of many actual accounting situations occurring in insurance companies. Nevertheless, it is valid for purposes of exposition and also can be adapted to suit the activities of any particular insurance firm.)
where

T = company tax rate;

λ_n = insurance premium loadings;

P_n = pure premiums,

X_n = claims;

R_n = investment returns;

PI_n = taxable profit/loss of sale of investments;

DR_n = dividends received,

DP_n = dividends paid;

E_n = all operating and administrative expenses;

PR_n = outward reinsurance premiums;

CR_n = reinsurance claims recovered.

In decision analysis the aim is to maximize expected utility of assets (A_D), i.e.

$$\text{Max } E[U(A_D)].$$

Since assets are a state variable, we can write:

$$A_{D,t} = A_{D,t-1} + \Delta A_D,$$

where

$$t = \text{time subscript.}$$

The original problem is therefore equivalent to:

$$\text{Max } E[U(\Delta A_D)].$$

$$\text{for a given } A_{D,t-1}.$$

In its current form the model does not address changes in capitalization (C) or reserves (R), and hence the system supports decisions whose objective is to maximize the expected utility of retained earnings [as defined in equation (3) above]. The model could be further generalized to account for changes in C or R.

ACKNOWLEDGEMENTS

The authors wish to thank Karen Fletcher of the University of Illinois and a number of anonymous reviewers for constructive comments and criticism. Versions of this paper were previously presented at a "Subjectivity in O.R." Symposium (London, October 1984) and at the I.I.F. Forecasting Conference (Montreal, June 1985).

REFERENCES

[1] R. L. Schultz and D. P. Slevin (Eds) (1975) *Implementing Operations Research Management Science.* Elsevier, New York.

[2] R. L. Ackoff (1979) The future of operational research is past. *J. Opl Res. Soc.* **30**, 93–104.

[3] R. L. Ackoff (1979) Resurrecting the future of operational research. *J. Opl Res. Soc.* **30**, 189–199.

[4] S. Eilon (1980) The role of management science. *J. Opl Res. Soc.* **31**, 17–28.

[5] H. C. Kunreuther and P. J. H. Schoemaker (1980) Decision analysis for complex systems integrating descriptive and prescriptive components. Working Paper, Department of Decision Sciences, University of Pennsylvania.

[6] R. D. Behn and J. W. Vaupel (1976) Why decision analysis is rarely used and how it can be. Working Paper Center for Policy Analysis, Institute of Policy Sciences and Public Affairs, Duke University.

[7] C. J. Grayson (1973) Management science and business practice. *Harv. Bus. Rev.* **51**, 41–48.

[8] R. V. Brown (1970) Do managers find decision theory useful? *Harv. Bus. Rev.* **48** 78–89.

[9] R. M. Hogarth (1980) *Judgement and Choice.* Wiley, New York.

[10] R. V. Brown, A. S. Kahr and C. R. Peterson (1974) *Decision Analysis for the Manager.* Holt Rinehart & Winston, New York.

[11] C. J. Grayson (1960) *Decisions Under Uncertainty: Oil and Gas Drilling Decisions.* Harvard University Press, Cambridge.

[12] P. G. Moore, H. Thomas, D. W. Bunn and J. M. Hampton (1976) *Case Studies in Decision Analysis.* Penguin Books, London.

[13] G. M. Kaufman and H. Thomas (1977) *Modern Decision Analysis.* Penguin Books, London.

[14] R. L. Keeney (1982) Decision analysis: state of the field. Technical Report 82-2, Woodward-Clyde Consultants, San Francisco. (Also published as "Decision analysis: an overview". *Opns Res.*).

[15] R. L. Keeney and H. Raiffa (1976) *Decisions with Multiple Objectives.* Wiley, New York.

[16] J. W. Ulvila and R. V. Brown (1982) Decision analysis comes of age. *Harv. Bus. Rev.* **60**, 130–141.

[17] R. A. Howard and J. E. Matheson (Eds) (1984) *Readings on the Principles and Applications of Decision Analysis*, Vols 1 and 2. Strategic Decisions Group, Menlo Park, Calif.

[18] M. B. McCaskey (1982) *The Executive Challenge.* Pitman, Marshfield, Mass.

[19] R. O. Mason and I. I. Mitroff (1981) *Challenging Strategic Planning Assumptions.* Wiley, New York.

[20] H. Raiffa (1968) *Decision Analysis.* Addison-Wesley, Reading, Mass.

[21] P. G. Moore and H. Thomas (1976) *The Anatomy of Decisions.* Penguin Books, London.

[22] R. L. Ackoff (1970) *A Concept of Corporate Planning.* Wiley, New York.

[23] R. A. Howard (1984) Value of information lotteries. In *Readings on the Principles and Applications of Decision Analysis* (R. A. Howard and J. E. Matheson, Eds). Strategic Decisions Group, Menlo Park, Calif.

[24] W. Edwards (1976) How to use multi-attributed measurement in social decision-making. Technical Report, 001597-1-T, Social Sciences Research Institute, University of Southern California.

[25] P. Humphreys and A. Wisudha (1979) Multi-attributed utility decomposition, MAUD. Technical Report, Decision Analysis Unit, Brunel University.

[26] G. T. Allison (1971) *Essence of Decision.* Little, Brown, Boston.

[27] A. Etzioni (1967) Mixed scanning: a third approach to decision-making. *Publ. Admin. Rev.* 385–391.

[28] R. L. Keeney and H. Raiffa (1972) A critique of formal analysis in public decision-making. In *Analysis of Public Systems* (A. W. Drake, R. L. Keeney and P. Morse, Eds), pp. 10–11. MIT Press, Cambridge.

[29] D. W. Bunn and H. Thomas (1977) Decision analysis and strategic policy formulation. *Long Range Plann.* **10**, 23–30.

[30] H. Thomas (1982) Screening policy options: an approach and a case study example. *Strateg. Mgmt J.* **3**, 277–244.

[31] C. W. Holling (1974) A case study of eco-system management. In *Project Status Report*, Ecology and Environment Projects, SR74-2, EC, IIASA, Austria.

[32] D. E. Bell (1977) A decision analysis of objectives for a forest pest problem. In *Conflicting Objectives in Decisions* (D. E. Bell, R. L. Keeney and H. Raiffa, Eds). Wiley New York.

[33] R. F. Meyer (1976) Preferences over time. In *Decisions with Multiple Objectives* (R. L. Keeney and H. Raiffa, Eds). Wiley, New York.

[34] D. B. Hertz and H. Thomas (1983) *Risk Analysis and its Applications.* Wiley, New York.

[35] D. B. Hertz and H. Thomas (1984) *Practical Risk Analysis.* Wiley, New York.

[36] C. R. Schwenk and H. Thomas (1983) Effects of conflicting analyses on managerial decision-making: a laboratory experiment. *Decis. Sci.* 467–482.
[37] C. W. Churchman (1971) *The Design of Inquiring Systems.* Basic Books, New York.
[38] R. O. Schlaifer (1971) *Computer Programs for Elementary Decision Analysis.* Harvard University Press, Cambridge.
[39] Stanford Research Institute (1976) Automated aids for analysis. S.R.I., Palo Alto.
[40] D. A. Samson and H. Thomas (1985) Decision analysis models in reinsurance. *Eur. J. Opl Res.* **19**, 201–211.
[41] C. R. Schwenk and H. Thomas (1983) Formulating the mess: the role of decisions aids in problem formulation. *Omega* **11**, 239–252.
[42] P. C. Humphreys (1980) Decision aids: aiding decisions. In *Decision Analysis and Decision Processes* (L. Sjoberg, T. Tyszka and J. A. Wise, Eds). Doya, Lund.
[43] P. Slovic (1980) Toward understanding and improving decisions. In *Human Performance and Productivity* (E. A. Fleishmann, Ed.).
[44] C. S. Spetzler and C. S. Stael Von Holstein (1984) Probability encoding in decision analysis. In *Readings on the Principles and Applications of Decision Analysis* (R. A. Howard and J. E. Matheson, Eds), pp. 601–620. Strategic Decisions Group. Menlo Park, Calif.
[45] P. G. Moore and H. Thomas (1975) Measuring uncertainty. *Omega.*
[46] B. Fischoff (1980) Clinical decision analysis, *Opns Res* 28–43.
[47] A. R. Lock and H. Thomas (1985) Making policy analysis palatable in organizations: the policy dialogue paradigm. Submitted for publication.
[48] H. C. Kunreuther (1982) The economics of protection against low probability events. *Analyzing and Aiding Decision Processes* (P. C. Humphreys and A. Vari, Eds). North-Holland, Amsterdam.
[49] H. Mintzberg, A. Raisinghani and D. Theoret (1976) The structure of 'unstructured' decision processes. *Admin. Sci. Q.* **21**, 246–275.
[50] M. Lyles and I. I. Mitroff (1980) On organizational problem forming: an empirical study. *Admin. Sci. Q.* **25**, 102–119.
[51] H. Jungermann (1980) Speculation about decision theoretic aids for personal decision making. *Acta Psychol.* **45**, 7–34.
[52] H. Jungermann and I. von Vlardt (1982) The role of the goal in representing decision problems. In *Analyzing and Aiding Decision Processes* (P. C. Humphreys and A. Vari, Eds). North-Holland, Amsterdam.
[53] L. D. Phillips (1984) Requisite decision modelling, *J. Opl. Res. Soc.* **33**, 303–312.
[54] H. Thomas (1984) Strategic decision analysis: applied decision analysis and its role in the strategic management: process. *Strateg. Mgmt J.* **5**, 139–156.
[55] G. M. Prince (1970) *The Practice of Creativity.* Macmillan, New York.
[56] C. R. Schwenk and R. A. Cosier (1980) Effects of the expert, devil's advocate and dialectical inquiry methods on prediction performance. *Org. Behav. Hum. Perform.* **26**, 409–424.
[57] W. C. Wedley, R. H. Jung and G. S. Merchant (1978) Problem solving the Delphi way. *J. gen. Mgmt* 23–36.
[58] L. I. Mitroff and J. R. Emshoff (1977) On strategy assumption making: a dialectical approach to policy and planning. *Acad. Mgmt Rev.* **4**, 1–12.
[59] L. W. Porter, E. E. Lawlgand J. (1975) *Behavior of Organizions.* McGraw-Hill, New York.
[60] K. Thomas (1976) Conflict and conflict management. In *Handbook of Industrial and Organizational Psychology* (M. D. Dunnette, Ed.). Rand-McNally, Chicago.
[61] K. R. MacCrimmon and R. W. Taylor (1970) Decision-making and problem solving.

In *Handbook of Industrial and Organizational Psychology* (M. D. Dunnette, Ed.). Rand-McNally, Chicago.

[62] D. B. Hertz and H. Thomas (1982) Evaluating the risks in acquisition. *Long Range Plann.* **15**, 38–44.

[63] A. R. Lock and H. Thomas (1985) Screening multi-attributed marketing strategy alternatives. In *Advances in strategic Management* (R. Lamb, Ed.), Vol. 4. JAI Press, Greenwich, Conn.

[64] D. A. Samson and H. Thomas (1983) Reinsurance decision-making and expected utility. *J. Risk Insur.* 249–264.

[65] H. Thomas (1983) Risk analysis and the formulation of diversification acquisition strategies. *Long Range Plann.* **16**, 28–38.

[66] H. Thomas (1985) Strategic management in research and development: a comparison between decisions in pharmaceuticals and electronics. *R&D Mgmt.* **15**, 3–22.

Further Reading

A full treatment of all the theory underpinning decision analysis may be found in my book *Decision Theory: An introduction to the mathematics of rationality* (French, 1986). In it I present and discuss all the theory that underlies Bayesian decision analysis. Moreover, I provide an extensive survey of the literature. It is, however, a book entirely on theory. True, I take considerable pains to interpret the theory and to relate it to applications; but I do not discuss any applications *per se.* (These notes and readings can be seen as a companion volume, providing the missing material on applications; although both projects were conceived entirely independently.) Smith (1987) also provides much of the theoretical background, with more emphasis on subjective probabilities and Bayesian statistics and less on utility theory and group decision making than my book. Keeney and Raiffa's book (1976) on multi-attribute utility is seminal: compulsory reading for all decision analysts.

Less theoretical, more applied approaches are taken by Bodily (1985), Bunn (1984), and von Winterfeldt and Edwards (1986). The special issue of the *Journal of the Operational Research Society* (1982, Vol. 33, No. 4) provides several case studies. For those who think they might not need decision analysis, Hogarth (1980), Kahneman *et al.* (1982) and Wright (1984) make salutary reading.

References

Adelson, R. M. (1965) Criteria for capital investment: an approach through decision theory, *Opl. Res. Q.*, **16**, 19–50.
(Extract reprinted in section 2.3.)

Adelson, R. M. and Norman, J. M. (1969) Operational research and decision-making, *Opl. Res. Q.*, **20**, 399–413.

Adelson, R. M. and Tocher, K. D. (1977) Viewpoints, *Opl. Res. Q.*, **28**, 106–109.
(Reprinted in section 4.1.)

Barnett, V. (1982) *Comparative Statistical Inference* (2nd edn), Wiley, Chichester.

Beattie, D. W. (1969) Marketing a new product, *Opl. Res. Q.*, **20**, 429–435.
(Reprinted in section 3.2.)

Beer, S. (1963) Review of C. W. Churchman's 'Prediction and optimal decision', *Opl. Res. Q.*, **14**, 343–350.

Belton, V. (1985) The use of a simple multi-criteria model to assist in selection from a shortlist, *J. Opl. Res. Soc.*, **36**, 265–274.
(Reprinted in section 3.7.)

Bodily, S. E. (1985) *Modern Decision Making: A guide to modelling with decision support systems*, McGraw-Hill, New York.

Bunn, D. W. (1984) *Applied Decision Analysis*, McGraw-Hill, New York.

Croston, J. D. and Gregory, G. (1969) A critique of Operational research and decision making' by Adelson and Norman, *Opl. Res. Q.*, **20**, 415–420.

DeGroot, M. H. (1970) *Optimal Statistical Decisions*, McGraw-Hill, New York.

Farquhar, P. H. (1981) Multivalent preference structures, *Math. Soc. Sci.*, **1**, 397–408.

Farquhar, P. H. (1984) Utility assessment methods, *Mgmt. Sci.*, **30**, 1283–1300.

Farquhar, P. H. and Fishburn, P. C. (1981) Equivalences and continuity in multivalent preference structures, *Ops. Res.*, **29**, 282–293.

French, S. (1983a) Decision analysis and life cycle costing. In J. K. Skwirzinsky (ed.), *Electronic Systems Effectiveness and Life Cycle Costing*, Springer-Verlag, Berlin, pp. 633–646.

French, S. (1983b) A survey and interpretation of multi-attribute utility. In S. French, R. Hartley, L. C. Thomas and D. J. White (eds), *Multi-objective Decision Making*, Academic Press, London, pp. 263–277.

French, S. (1984) From decision theory to decision analysis. In R. W. Eglese and G. K. Rand (eds), *Developments in Operational Research*, Pergamon Press, Oxford, pp. 77–87.
(Reprinted in section 4.3.)

French, S. (1985) Group consensus probability distributions: a critical survey. In J. M. Bernardo, M. H. DeGroot. D. V. Lindley and A. F. M. Smith (eds), *Bayesian Statistics II*, North-Holland, Amsterdam, pp. 183–202.

French, S. (1986) *Decision Theory: An introduction to the mathematics of rationality*, Ellis Horwood, Chichester.

French, S., Hartley, R., Thomas, L. C., White, D. J. and Tocher, K. D. (1978)

Viewpoints, *J. Opl. Res. Soc.*, **29**, 1132–1135.
(Reprinted in section 4.1.)

French, S. and Tocher, K. D. (1978) Viewpoints, *J. Opl. Res. Soc.*, **29**, 179–182.
(Reprinted in section 4.1.)

Hall, P. (1986) Managing change and gaining corporate commitment, *ICL Technical Journal*, **7**, 213–227.

Hampton, J. M., Moore, P. G. and Thomas, H. (1973) Subjective probability and its measurement, *J. Roy. Statist. Soc.*, A, **136**, 21–42.

Hogarth, R. (1980) *Judgement and Choice*, Wiley, Chichester.

Hull, J. C., Moore, P. G. and Thomas, H. (1973) Utility and its measurement, *J. Roy. Statist. Soc.*, A, **136**, 226–247.

Humphreys, P. C. and Wisudha, A. (1987) Methods and tools for structuring and analysing decision problems: Volumes 1 and 2, *Technical Report* 87-1, Decision Analysis Unit, London School of Economics and Political Science.

Kahneman, D., Slovic, P. and Tversky, A. (eds) (1982) *Judgement Under Uncertainty: Heuristics and biases*, Cambridge University Press.

Keefer, D. L. and Kirkwood, C. W. (1978) A multi-objective decision analysis: budget planning for product engineering, *J. Opl. Res. Soc.*, **29**, 435–442.
(Reprinted in section 3.4.)

Keeney, R. L. and Raiffa, H. (1976) *Decisions with Multiple Objectives: Preferences and value trade-offs*, Wiley, New York.

Kirkwood, C. W. (1982) A case history of nuclear power plant site selection, *J. Opl. Res. Soc.*, **33**, 353–363.

Kyburg, H. E., jr. (1970) *Probability and Inductive Logic*, Macmillan, London.

Lathrop, J. W. and Watson, S. R. (1982) Decision analysis for the evaluation of risk in nuclear waste management, *J. Opl. Res. Soc.*, **33**, 407–418.
(Reprinted in section 3.5.)

Lindley, D. V. (1985) *Making Decisions* (2nd edn), Wiley, Chichester.

McCord, M. and de Neufville, R. (1983) Fundamental deficiency of expected utility decision analysis. In S. French, R. Hartley, L. C. Thomas and D. J. White (eds), *Multi-objective Decision Making*, Academic Press, London, pp. 279–305.

Merkhofer, M. W. (1987) Quantifying judgemental uncertainty: methodology, experiences and insights, *IEEE Trans. on Systems, Man and Cybernetics*, **17**, 741–752.

Moore, P. G. and Thomas, H. (1973) The rev counter decision, *Opl. Res. Q.*, **24**, 337–351.
(Reprinted in section 1.2.)

Pearman, A. D. (1987) The application of decision analysis: a US/UK comparison, *J. Opl. Res. Soc.*, **38**, 775–783.

Phillips, L. D. (1982) Requisite decision modelling: a case study, *J. Opl. Res. Soc.*, **33**, 303–311.
(Reprinted in section 3.6.)

Phillips, L. D. (1984) A theory of requisite decision models, *Acta Psych.*, **56**, 29–48.

Raiffa, H. (1968) *Decision Analysis*, Addison-Wesley, Reading, Mass.

Ravinder, H. V., Kleinmuntz, D. N. and Dyer, J. S. (1988) The reliability of subjective probabilities obtained through decomposition, *Mgmt. Sci.*, **34**, 186–199.

Ronen, B., Pliskin, J. S. and Feldman, S. (1984) Balancing the failure modes in the electronic circuit of a cardiac pacemaker: a decision analysis, *J. Opl. Res. Soc.*, **35**, 379–387.
(Reprinted in section 3.3.)

Savage, L. J. (1971) Elicitation of personal probabilities and expectations, *J. Amer. Statist. Assoc.*, **66**, 783–801.

Savage, L. J. (1972) *The Foundations of Statistics* (2nd edn), Dover, New York.

Spetzler, C. S. and Stael von Holstein, C. S. (1975) Probability encoding in decision analysis, *Mgmt. Sci.*, **22**, 340–358.

Smith, J. Q. (1987) *Decision Analysis: A Bayesian approach*, Chapman and Hall, London.

Thomas, H. and Samson, D. (1986) Subjective aspects of the art of decision analysis: exploring the role of decision analysis in decision structuring, decision support and policy dialogue, *J. Opl. Res. Soc.*, **37**, 249–265.
(Reprinted in section 4.5.)

Tocher, K. D. (1976) Notes for discussion on 'Control', *Opl. Res. Q.*, **27**, 231–239.
(Extract reprinted in section 4.1.)

von Winterfeldt, D. and Edwards, W. (1986) *Decision Analysis and Behavioural Research*, Cambridge University Press.

Wallsten, T. S. and Budescu, D. V. (1983) Encoding subjective probabilities: a psychological and psychometric review, *Mgmt. Sci.*, **29**, 151–174.

Wells, G. E. (1982) The use of decision analysis in Imperial Group, *J. Opl. Res. Soc.*, **33**, 313–318.

White, D. J. (1976) *Decision Methodology*, Wiley, Chichester.

Wright, G. (1984) *Behavioural Decision Theory*, Penguin Books, Harmondsworth.

Index